PETROCHEMICALS

UMIST SERIES IN SCIENCE AND TECHNOLOGY
This series establishes a modern and up-to-date range of teaching books for sixth-formers and first-year undergraduates or the equivalent in other establishments in tertiary education, including polytechnics, colleges of further education and teacher training colleges. The areas covered, in the form of short accounts of particular subjects, will span a range of disciplines in the pure and applied sciences and in engineering, drawing upon the expertise and knowledge of staff within the University of Manchester Institute of Science and Technology.

Editorial Panel:
Professor B. Launder, Dr. R. V. Parish and Ann Buckley, B.Sc. (all of UMIST)

PETROCHEMICALS
P. Wiseman, UMIST

PETROCHEMICALS

P. WISEMAN, B.Sc.(Hons.)
Department of Chemistry
University of Manchester Institute of Science and Technology

ELLIS HORWOOD LIMITED
Publishers · Chichester

Halsted Press: a division of
JOHN WILEY & SONS
New York · Chichester · Brisbane · Toronto

First published in 1986 by
ELLIS HORWOOD LIMITED
Market Cross House, Cooper Street, Chichester, West Sussex, PO19 1EB,
England
in association with
**UNIVERSITY OF MANCHESTER INSTITUTE OF SCIENCE AND
TECHNOLOGY**

Distributors:

Australia, New Zealand, South-east Asia:
Jacaranda-Wiley Ltd., Jacaranda Press,
JOHN WILEY & SONS INC.,
G.P.O. Box 859, Brisbane, Queensland 4001, Australia

Canada:
JOHN WILEY & SONS CANADA LIMITED
22 Worcester Road, Rexdale, Ontario, Canada.

Europe, Africa:
JOHN WILEY & SONS LIMITED
Baffins Lane, Chichester, West Sussex, England.

North and South America and the rest of the world:
Halsted Press: a division of
JOHN WILEY & SONS
605 Third Avenue, New York, N.Y. 10158 U.S.A.

© **1986 P. Wiseman/Ellis Horwood Limited**

British Library Cataloguing in Publication Data
Wiseman, P.
Petrochemicals. —
(UMIST Series in Science & Technology)
1. Petroleum chemicals
I. Title
661'.804   TP692.3

Library of Congress Card No. 85–24843

ISBN 0–85312–741–7 (Ellis Horwood Limited — Library Edn.)
ISBN 0–85312–978–9 (Ellis Horwood Limited — Student Edn.)
ISBN 0–470–20279–3 (Halsted Press — Library Edn.)
ISBN 0–470–20284–X (Halsted Press — Student Edn.)

Typeset by Ellis Horwood Limited
Printed in Great Britain by Unwin Brothers of Woking

Table of Contents

Preface 11

Chapter 1 — Introduction
   1.1 What are petrochemicals? 13
   1.3 The development of the petrochemical industry 14
   1.3 The raw materials — petroleum and natural gas 15
   1.4 Petroleum refining. 17
   1.5 Petrochemical feedstocks from petroleum refining 20
   1.6 The basic building block processes 20
   1.7 Petrochemical process technology. 21
      1.7.1 Reactors 23
      1.7.2 Product separation 24
   1.8 Costs in chemical processing 26
      1.8.1   Chemistry and costs. 27
      1.8.2 Costs and scale of operation 28
   1.9 Nomenclature 29
   Problems and exercises 29

Chapter 2 — Ethylene and Co-products by Thermal Cracking
   2.1 Feedstocks 30
   2.2 The reaction system 31

2.3   The primary cracking reactions. 34
   2.3.1   Ethane cracking 34
   2.3.2   Propane cracking. 35
   2.3.3   Naphtha and gas oil cracking 36
2.4   Secondary reactions 39
2.5   Products and product separation 41
Problems and exercises 42

Chapter 3 — Ethylene Derivatives

3.1   Polyethylene 44
3.2   Vinyl chloride 47
   3.2.1   Uses of vinyl chloride 49
3.3   Ethylene oxide and derivatives. 49
   3.3.1   Ethylene glycol 51
   3.3.2   Ethanolamines 52
   3.3.3   Polyoxyethylene derivatives 53
3.4   Styrene 53
3.5   Ethanol 53
3.6   Acetaldehyde 55
   3.6.1   Uses of acetaldehyde 58
3.7   Vinyl acetate 59
3.8   Chlorinated solvents 60
3.9   Ethyl chloride 60
3.10 Linear 1-alkenes and linear primary alcohols 61
Problems and exercises 63

Chapter 4 — Propylene Derivatives

4.1   Polypropylene 65
4.2   Isopropyl alcohol and acetone 66
4.3   Propylene oxide 68
   4.3.1   Uses of propylene oxide 71
4.4   Cumene 72
4.5   Hydroformylation: butyl alcohol and other products. . . . 72
   4.5.1   Conventional Oxo process. 72
   4.5.2   Newer proceses 73
   4.5.3   Hydroformylation products 74
4.6   Propylene trimer and tetramer 74
4.7   Acrylic acid and acrylates 75
4.8   Acrylonitrile 76
4.9   Allyl chloride 77
Problems and exercises 78

Chapter 5 — Butadiene and Butenes
  5.1 Introduction 80
  5.2 Butadiene — manufacture 81
  5.3 Butadiene — uses 82
    5.3.1 Synthetic rubbers 82
    5.3.2 Chloroprene 83
    5.3.3 Adiponitrile 84
  5.4 Isobutane 85
  5.5 n-Butenes 87
  Problems and exercises 87

Chapter 6 — Benzene, Toluene and Xylenes Production
  6.1 Introduction 89
  6.2 Catalytic reforming 90
    6.2.1 Chemistry of reforming 90
    6.2.2 Reaction systems 93
    6.2.3 Product separation 93
  6.3 Aromatics from pyrolysis gasoline 95
  6.4 C_8 aromatics 96
    6.4.1 Separation 97
    6.4.2 Isomerisation 98
  6.5 Hydrodealkylation and disproportionation 99
  Problems and exercises 100

Chapter 7 — Benzene Derivatives
  7.1 Ethylbenzene and styrene 101
    7.1.1 Ethylation 102
    7.1.2 Dehydrogenation 104
    7.1.3 Uses of styrene 104
  7.2 Cyclohexane and nylon intermediates 105
    7.2.1 Cyclohexane 105
    7.2.2 Cyclohexanol and cyclohexanone 106
    7.2.3 Adipic acid 109
    7.2.4 Adiponitrile and hexamethylenediamine 109
    7.2.5 Nylon 66 110
    7.2.6 Caprolactam 111
    7.2.7 Nylon 6 112
  7.3 Phenol 113
    7.3.1 Routes to phenol 113
    7.3.2 The cumene process 115
    7.3.3 Uses of phenol 117

7.4   Nitrobenzene and aniline 120
7.5   Maleic anhydride 122
7.6   Detergent alkylate and alkylbenzenesulphonate detergents   123
Problems and exercises 124

Chapter 8 — Toluene and Xylene Derivatives
8.1   Tolylene di-isocyanate 126
8.2   Benzoic acid. 129
8.3   Phthalic anhydride. 132
8.4.1   Uses of phthalic anhydride. 133
8.5   Terephthalic acid and dimethyl terephthalate 134
8.5.1   Poly(ethylene terephthalate) 138
Problems and exercises 139

Chapter 9 — Steam Reforming and Related Processes
9.1   The reforming reactions 141
9.2   Steam reforming for ammonia 143
9.3   Ammonia synthesis 146
9.4   Uses of ammonia 147
9.5   Methanol — manufacture 148
9.6   Methanol — uses and derivatives 150
9.6.1   Formaldehyde 150
9.6.2   Acetic acid 151
9.6.3   Methylamines 152
9.6.4   Methyl chloride 153
9.6.5   Other uses of methanol 153
9.7   Steam reforming for other purposes 155
9.8   Synthesis gas by other methods 156
9.8.1   From coal 156
9.8.2   From petroleum fractions by partial oxidation . . . 157
Problems and exercises 158

Chapter 10 — Miscellaneous Processes and Products
10.1   Chloromethanes 159
10.2   Carbon disulphide 162
10.3   Acetic acid by butane and naphtha oxidation 163
10.4   Dehydrogenation and chlorination of linear alkanes . . . 164
10.5   Wax cracking 165
10.6   Cracking for acetylene 166

Problems and exercises 167

Sources of Further Information and Further Reading 169

Answers to Problems 172

Index 177

Preface

Petrochemicals can lay two main claims to the attention of those who are interested in chemistry. Firstly, the petrochemical industry is a major industry of great importance in modern society and provides a huge range of useful products, such as plastics, rubbers, fibres, detergents, and solvents, and supplies raw materials for the manufacture of many other chemicals. Secondly, it has a coherent structure made up of an interlocking system of raw materials, processes, and products, which makes it an ideal medium for observing factors that influence the application of chemistry in industry. In this book I have attempted to bring out both of these aspects. Whilst it is written primarily for students and teachers of chemistry and chemical engineering, it should also be useful to practising chemists and chemical engineers as a concise account of what goes on in the petrochemical industry.

The book is based on undergraduate and postgraduate courses given at UMIST for a number of years. After an introductory chapter, the text is structured mainly round the three basic building block petrochemical processes, cracking, catalytic reforming and steam reforming, and the downstream processing of the intermediates that they produce. Reaction mechanisms are discussed where particular points of interest arise, but no attempt is made at a comprehensive discussion of mechanisms: this would have extended the book far beyond its intended length. Throughout, relationships between process chemistry, technology, and economics have been emphasised. To help develop a feeling for the scale of petrochemical operations, details of US capacities are given for major products. (US data are used because they are much more readily available than those for other countries. As a rough rule of thumb, the total West European petrochemical industry is about the same size as the US industry.) Problems are provided at the end of each chapter; students are strongly recommended to attempt these.

When a book is based on teaching and researches that have been carried out over a substantial period it is inevitable that many people will have made inputs of one sort or another, and it is impossible to acknowledge or even identify all of these. However, I do wish to record my debt to the following students who have carried out research projects on the petrochemical industry and related topics under my supervision, and whose work provided stimulation and information: S. M. Abdul-Hadi, O. M. J. Adams, Zed Alatas, Harout Bronozian, P. J. Campbell, C. Y. Cheung, O. A. Garcia, J. E. Hartley, S. S. Islam, I. P. Kohli, P. A. Obuasi, A. B. Summerscales, and Farouk Warris. To any of these who happen to read this book, best wishes and thanks. I would also like to express my thanks to the series editor Dr R. V. Parish for his encouragement and many helpful comments.

Finally, I gratefully acknowledge the invaluable help given by my wife Hilary, both in general support and encouragement and in her professional assistance with the subtleties of the English language.

P. Wiseman
Manchester, July 1985

1

Introduction

1.1 WHAT ARE PETROCHEMICALS?

In a sense, most organic chemical products are petrochemicals. If you take an aspirin tablet it is highly likely that the carbon atoms and some of the hydrogen atoms it contains came originally from petroleum and possibly natural gas. However, the use of the term 'petrochemical' is usually confined to products that are derived from petroleum and natural gas *and* are made on a large scale. Phenol and acetic anhydride are petrochemicals; aspirin, which is derived from them, is not. It is not possible to define precisely what is meant by 'a large scale', but roughly it means products which are made in plants of capacities from about 10 000 tonnes per annum upwards. Most of the processes discussed in this book are operated in plants with capacities in the range 20 000 to 650 000 tonnes per annum.

In terms of tonnage, about 95% of total organic chemical manufacture is at present based on petroleum and natural gas. It is, however, important not to overestimate the importance of the petrochemical industry as a consumer of these materials. In the USA, for instance, only about 5% of petroleum is used as raw material for chemical manufacture. Its main use is as a fuel, as is the main use of natural gas.

It is commonly believed that petrochemicals are made by the oil industry − a 'by-product of refining'. This is a misconception.

The petrochemical industry is a part of the chemical industry, and quite distinct from refining, except in so far as it uses some products of refining as raw materials. Many companies that make petrochemicals do not carry out oil refining, and those oil companies which make petrochemicals tend to have a separate organisation for this part of the business. The oil industry's main business is fuels, with raw materials for petrochemicals as an important side-line. The petrochemical industry's main business is chemicals.

1.2 THE DEVELOPMENT OF THE PETROCHEMICAL INDUSTRY

It is generally agreed that the development of the industry started in 1920, when the Standard Oil Company of New Jersey USA brought a plant into operation for the manufacture of isopropyl alcohol from propylene. The propylene used in this process was a by-product of gasoline production by cracking, a process which had been introduced into the oil industry in 1913. The first petrochemical manufacture based on ethylene was in 1923, when Union Carbide started the production of ethylene chlorohydrin, ethylene glycol and dichloroethane at South Charleston, West Virginia, based on ethylene made by cracking ethane. The industry developed steadily in the USA during the 1920s and 1930s, but was then given an immense stimulus by World War II. By 1945 the petrochemical industry was more or less fully developed in the USA.

Outside the USA, petrochemical manufacture was carried out to only a very limited extent before the 1950s. During that decade petrochemical industries built up, first in the UK and then throughout Western Europe. The 1960s saw the spread of petrochemical manufacture to all developed countries, and massive developments in scale of operation and in process technology. By the end of this decade the petrochemical industry was established on the world scale.

The oil crisis of 1973 and the huge increases in the price of oil which followed have had major effects on the industry, and these are still developing. The rate of growth of the petrochemical industry, and of the chemical industry in general, is now much less than it was in the golden days of the 1950s and the 1960s. The technology is changing in response to the increases in price of hydrocarbon raw materials, and of fuel. Possibly the most important effect in the next few years will be a small but significant shift in the geographical centre of gravity of the industry. Many of the oil-producing developing countries are at present setting up major petrochemical developments based on gas, from oil fields, which would otherwise be wasted. Saudi Arabia, for example, is currently bringing into operation plants which will be able to supply about 3% of total world requirements for ethylene and ethylene derivatives.

1.3 THE RAW MATERIALS – PETROLEUM AND NATURAL GAS

Petroleum, or **crude oil**, occurs underground in reservoirs in porous rock in certain types of geological structures. The oil is recovered from these by drilling oil wells from which it either flows out under the reservoir pressure or is pumped out.

Major deposits of oil occur in only a fairly limited number of regions (see Table 1.1). A notable feature is that more than half the world's known reserves are in the Middle East. Many heavily industrialised countries have only very small reserves of oil. Even the USA, the pioneer of the oil industry and one of the world's major oil producers has to import about half its requirements.

Table 1.1
Distribution of oil and natural gas reserves (%)

	Oil	Natural gas
Africa	9.8	5.9
Asia and Australasia	6.3	5.8
Middle East	54.0	24.2
Latin America	9.5	5.8
North America	7.3	9.1
USSR and Eastern Europe	10.3	44.3
Western Europe	2.8	4.9

Source of data: *Oil and Gas Journal*, **83**, 52, 80–81.

Crude oils vary widely from region to region and from field to field. Some crudes are light-coloured, mobile liquids, others are thick, treacly or tarry materials. Some have quite pleasant smells, others smell quite disgusting. However, all are made up mainly of three types of hydrocarbons, alkanes, cycloalkanes, and aromatics. The differences in character arise from the different proportions of hydrocarbon types, differences in molecular weight range, and differences in the nature and amount of non-hydrocarbon materials.

The alkanes, called *paraffins* in the petroleum industry, range from methane upwards. Both straight chain and branched chain alkanes are present, in varying proportions depending on the type of crude. The cycloalkanes, called *naphthenes* in the industry, contain five and six membered ring compounds only. The aromatics range from benzene upwards. The proportions of these different types of structures vary over a wide range. Aromatics commonly make up the smallest proportion of the oil.

The molecular weight range covered varies greatly. Oils containing predominantly low molecular weight components are called 'light' oils; those containing a high proportion of high molecular weight materials are called 'heavy' oils.

Oils contain small percentages of sulphur-, oxygen-, and nitrogen-containing organic compounds, and traces of metallic compounds, notably of vanadium.

There are two sources of **natural gas**. It occurs in gas fields, underground reservoirs broadly similar to oil reservoirs, from which it is recovered by drilling gas wells. In addition, large amounts of gas are produced in association with the production of crude oil. We have already seen that oil contains alkanes from methane upwards. In the reservoir the lower gaseous alkanes are often in solution under considerable pressure. When the oil is brought to the surface the pressure is released and a proportion of the C_1 to C_4 alkanes boils off. This is called *associated gas*. In some oil fields, e.g. those in inaccessible regions, this gas is simply burnt. In others it is collected and used.

Natural gas varies widely in composition. Some gases contain only small amounts of alkanes other than methane. These are called *dry natural gas*. This term derives from the fact that if such a gas is compressed at ambient temperature no appreciable amount of liquid is formed. Other gases contain substantial amounts of ethane, propane and butanes, which liquefy if the gas is compressed. These are called *wet natural gas*. Associated gas is invariably wet, whereas gas from gas fields is usually dry. In addition to hydrocarbons, natural gas sometimes contains substantial proportions of other components such as carbon dioxide, nitrogen, hydrogen sulphide and helium. Table 1.2 shows some examples of natural gas compositions.

Dry natural gas is generally used as it stands, except that hydrogen sulphide, if present, would usually be removed, as would most of the carbon dioxide if initially present in large amounts. Wet natural gas is often stripped of propane and butane and most of the ethane, either by scrubbing with a hydrocarbon solvent at about −45°C, or by cooling to about −120°C so that the ethane, propane and butanes liquefy. The ethane, propane, and butanes may then be separated by low temperature distillation as required.

Table 1.2

Composition of selected natural gases (% by volume)

NON-ASSOCIATED

Area:	Algeria	France	Holland	New Zealand	North Sea
Field:	Hassi-R'Mel	Lacq	Groningen	Kapuni	West Sole
CH_4	83.5	69.3	81.3	46.2	94.4
C_2H_6	7.0	3.1	2.9	5.2	3.1
C_3H_8	2.0	1.1	0.4	2.0	0.5
C_4H_{10}	0.8	0.6	0.1	0.6	0.2
C_5^+	0.4	0.7	0.1	0.1	0.2
N_2	6.1	0.4	14.3	1.0	1.1
CO_2	0.2	9.6	0.9	44.9	0.5
H_2S	–	15.2	trace	–	–

ASSOCIATED

Area:	Abu Dhabi	Iran	North Sea	North Sea
Field:	Zakum	Agha Jari	Forties	Brent
CH_4	76.0	66.0	44.5	82.0
C_2H_6	11.4	14.0	13.3	9.4
C_3H_8	5.4	10.5	20.8	4.7
C_4H_{10}	2.2	5.0	11.1	1.6
C_5^+	1.3	2.0	8.4	0.7
N_2	1.1	1.0	1.3	0.9
CO_2	2.3	1.5	0.6	0.7
H_2S	0.3	–	–	–

From *Our Industry Petroleum*, 5th edn, London, The British Petroleum Company Ltd. Courtesy British Petroleum.

1.4 PETROLEUM REFINING

Crude oil is not used as such, either as a fuel, or as a chemical feedstock. It is first refined.

The first major operation in refining is fractional distillation of the crude oil to separate it into fractions with boiling ranges appropriate to the major fuel applications. Typical fractions taken are indicated in Table 1.3. It can be seen from the table that one of the fractions, naphtha, has no direct use as a fuel. Before we discuss what is done with the naphtha, it is necessary to understand something of the quality requirements for one of the major refinery products, gasoline,† the fuel used in spark ignition internal combustion engines.

Table 1.3
Fractions from distillation of crude oil

Fraction	Boiling point (°C)	Fuel applications
Gases	<20	Refinery fuel Liquefied petroleum gas
Light gasoline	20–75	Gasoline blending
Naphtha	75–200	
Kerosine	200–250	Jet fuel Domestic fuel Tractor fuel
Gas oil	250–350	Diesel fuel Heating fuel
Atmospheric residue	>350	Heavy fuel oil

Note: This presents a simplified picture. The boiling ranges may be varied somewhat, and will in any case overlap. Extra fractions may be taken.

There are two major quality requirements for gasoline. Firstly, it has to have the correct volatility characteristics, so that it vaporises adequately in the carburettor, but does not undergo excessive evaporation from the fuel tank, or give rise to 'vapour locking' in the fuel pipes. A fraction with a boiling range of about 20 to 200°C gives the best compromise between these conflicting requirements.

The other requirement is that the gasoline should have the correct burning characteristics. In particular, it should not give rise to 'knocking' or 'pinking', since this results in power loss and, if severe, to engine damage. The *octane number* of a fuel is a measure of its quality in this respect, the higher the octane number, the less the tendency of the fuel to knock. Most modern car engines require fuels with octane numbers in the range 90 to 98.

† In the UK, gasoline is called petrol. This term is often confused with petroleum and so we shall avoid it. Rather, the American, and international, term will be used.

A fraction of boiling range 20 to 200°C taken from a typical crude oil would have an octane number of only about 50, and would be totally unsuitable for use in modern cars. In fact, as can be seen from Table 1.3, two fractions are taken off in this range. The light gasoline has an octane number of about 65, and, provided that it is blended with other components of high octane number, can be used in motor gasoline without further treatment. The naphtha has an octane number of only about 40, and has to be chemically processed before it can be used in gasoline. The process used is **catalytic reforming**. This partially converts alkanes and cycloalkanes into aromatic hydrocarbons, which have very high octane numbers, and gives *catalytic reformate*, with an octane number of 95 to 100. Catalytic reforming is also used as a petrochemical process, for the manufacture of aromatic hydrocarbons, and we shall discuss it further in Chapter 6.

Satisfactory motor fuel can be made by blending light gasoline and catalytic reformate, together with various additives. However, operation in this way would in many circumstances produce too little gasoline and too much heavy fuel oil. Consequently, refineries operate processes to balance the output of fuel products with market requirements. The most important of these is **catalytic cracking**.

Catalytic cracking is usually operated on *vacuum distillate*, produced by distilling atmospheric residue under vacuum. This is contacted with a strongly acidic solid catalyst at about 500°C. Molecular weight reduction reactions occur to produce gasoline, with an octane number of about 90, together with higher boiling fractions and catalytic cracker gas. This gas, which is formed to the extent of 10% to 20% of the feed, contains substantial amounts of propylene and butenes (see Table 1.4).

Table 1.4
Catalytic cracker gases — typical composition

	wt %
H_2	0.6
H_2S	3.4
CH_4	10.3
C_2H_4	3.4
C_2H_6	7.5
C_3H_6	13.8
C_3H_8	12.1
C_4H_8	18.4
C_4H_{10}	30.5

1.5 PETROCHEMICAL FEEDSTOCKS FROM PETROLEUM REFINING

The refinery products of most importance as petrochemical feedstocks are naphtha, gas oil, and catalytic cracker gases.

In the case of naphtha, we have to live with an unfortunate confusion of terminology. The term has somewhat different meanings in the oil industry and the petrochemical industry. In the petrochemical industry, any petroleum fraction in the approximate boiling range 20 to 200°C which is used as a feedstock is called naphtha. Thus, if the fraction called light gasoline in Table 1.3 was being used as feedstock for ethylene manufacture, it would be called light naphtha. Henceforth, when the term naphtha is used, it will be in the petrochemical sense.

Naphthas are mixtures of alkanes, cycloalkanes, and aromatic hydrocarbons in proportions which vary considerably depending on the type of oil from which the naphtha was derived. A 'full range' naphtha, boiling range approximately 20 to 200°C, would contain compounds with from 4 to 12 carbon atoms.

Gas oils are also mixtures of alkanes, cycloalkanes, and aromatic hydrocarbons, but the components have higher molecular weights than those of naphtha. Gas oil can be used in place of naphtha as a feedstock for ethylene manufacture but is somewhat less technologically convenient for this purpose. Consequently, it is only economically attractive when it is substantially cheaper than naphtha.

In the USA, where catalytic cracking is carried out on an extremely large scale, catalytic cracker gases provide the major source of propylene and butenes. In Western Europe and Japan, catalytic cracking is carried out on a much smaller scale, and catalytic cracker gases are much less important as a feedstock. Catalytic cracker gases are *not* important as a source of ethylene since the proportion that they contain is too low for its separation to be worthwhile.

Kerosine is used to some extent as a chemical feedstock, but only in specialised applications. It could be used for the manufacture of ethylene, but is economically unattractive for this since its price, which is determined by its value as a fuel, is relatively high.

The gaseous alkanes produced in crude oil distillation and in catalytic cracking are sometimes used as feedstocks, but their utility is limited by the fact that the amounts available, after demands for refinery fuel and liquefied petroleum gas have been met, are relatively small.

1.6 THE BASIC BUILDING BLOCK PROCESSES

A large proportion of petrochemical manufacture is based on intermediates made by three basic building block processes, *thermal cracking, catalytic reforming*, and *steam reforming*. These processes, and downstream processes based on the intermediates, are discussed in detail in Chapters 2 to 9. However, at this stage it will be useful to consider them in outline, to provide an overall view of the technological structure of the industry.

Thermal cracking, also known as steam cracking, is used for the manufacture of ethylene. The feedstocks commonly used are ethane, propane, naphtha and gas oil, depending on availability and price. Typically, ethane and propane are the predominant feedstocks in regions where substantial amounts of wet natural gas are available. When ethane is cracked, ethylene is the only major product. Propane gives substantial amounts of propylene as co-product, and naphtha and gas oil give propylene, butenes and butadiene, and aromatic hydrocarbons. All these intermediates may be converted into a number of products, so that ethylene plants tend to form the centre of large petrochemical complexes.

Catalytic reforming, which we have already briefly met as a refining process, is used in the petrochemical industry for making benzene, toluene, and xylenes. The feedstock used is naphtha. Benzene has a wide range of derivatives, but toluene and the xylenes have much more restricted applications. In regions where naphtha or gas oil are major feedstocks for ethylene production, ethylene plants are also important producers of aromatic hydrocarbons.

Steam reforming produces, initially, a mixture of carbon monoxide and hydrogen, which may subsequently be used for a variety of purposes. The two most important applications are in making ammonia and methanol. Where natural gas is available, methane is the usual feedstock. Otherwise, naphtha is used.

1.7 PETROCHEMICAL PROCESS TECHNOLOGY

Petrochemicals are usually made by *continuous processing*, that is, raw materials are continuously fed into the plant, and products are continuously taken from it. This contrasts with most laboratory preparations which are carried out by *batch processing*. Continuous processing is used because it has a number of advantages for large-scale chemical manufacture, notably, it is easier to automate, it allows better energy conservation, and it allows short reaction times to be used.

Typically, a petrochemical plant will have one or more reaction systems, depending on how many separate reaction stages there are in the process, and one or more separation systems. Let us take a specific example. Styrene is made in two stages from benzene and ethylene:

$$C_6H_6 + C_2H_4 \longrightarrow C_6H_5C_2H_5 \longrightarrow C_6H_5CH=CH_2 + H_2$$

In the first stage an excess of benzene is used to reduce the extent of polyalkylation; the product stream from this stage contains ethylbenzene, unreacted benzene, and polyethylbenzenes. The ethylbenzene is separated and dehydrogenated. Some benzene and toluene are formed in this reaction, and not all the benzene is converted, so the product stream consists of styrene, ethylbenzene, benzene, toluene, and hydrogen. This is separated into its components, and the ethylbenzene and benzene are recycled. The process is conveniently represented by an *outline flow scheme*, as shown in Fig. 1.1.

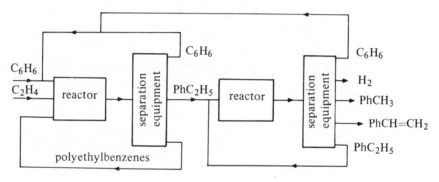

Fig. 1.1 – Outline flow scheme, styrene process.

In both stages of this process a raw material is only partially converted, and is *recycled* to the reactor. This is very common in petrochemical processes. In discussing processes involving recycling it is important to understand precisely what is meant by the term 'yield', particularly since there is some variation in the way in which it is used. In this book, when the term yield is used in connection with a process, or stage of a process, in which there is recycling of one or more raw materials, unless otherwise specified it means the yield based on raw material consumed by the stage after allowing for recycling. An example will help to make this clear. Suppose that the dehydrogenation stage of a styrene plant is operating as indicated in Fig. 1.2. Although the feed rate of ethylbenzene to the reactor is 100 moles sec^{-1}, the rate of consumption of ethylbenzene by the stage is only 40 moles sec^{-1}, since 60 moles sec^{-1} is being recycled. The yield over the stage is therefore:

$$\frac{36}{40} \times 100 = 90\%$$

The *conversion* in this stage, that is the percentage of raw material consumed in the reaction, is 40%.

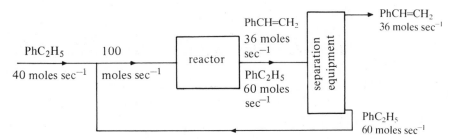

Fig. 1.2 — Dehydrogenation stage, styrene process. (For simplicity, 100% recovery has been assumed in the separation equipment. Also, hydrogen, benzene and toluene are not shown.)

1.7.1 Reactors

There are two main types of reactors used for carrying out reactions continuously in the liquid phase, *tubular reactors*, and *stirred flow reactors* (see Fig. 1.3). Tubular reactors consist, as the name implies,

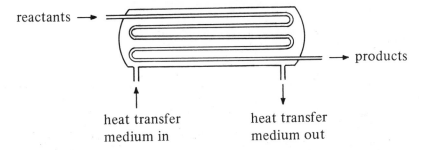

tubular reactor

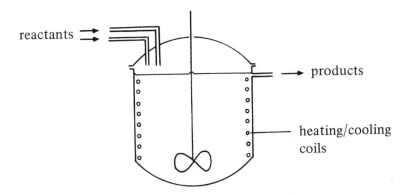

stirred flow reactor

Fig. 1.3 — Reactors for continuous processing in the liquid phase.

of a tube, or tubes, through which the reaction mix flows. Usually, provision is made for the addition or removal of heat by a heat transfer medium circulating round the tubes. A stirred flow reactor consists of a vessel with an agitator and provision for continuously adding reactants and removing products. Heat transfer may be provided for by a variety of means, e.g. by a jacket, or by coils immersed in the reaction mix.

The type of reactor used for gas phase processes depends on whether the reaction involved is carried out in contact with a solid catalyst or not. Non-catalytic gas phase reactions are carried out in tubular reactors broadly similar to those used for liquid phase reactions. With reactions over solid catalysts, considerations of heat transfer become dominant. Beds of catalyst are poor transmitters of heat, so that there is a tendency for large temperature gradients to be set up.

In cases where the heat of reaction is low, or substantial temperature gradients can be tolerated, *adiabatic reactors* can be used. In these, no provision for heat transfer is made, the catalyst simply being held as a bed in a suitable vessel (see Fig. 1.4(a)). The reactants are preheated to an appropriate temperature, and are passed into the reactor. The temperature rises or falls through the bed depending on whether the reaction is exothermic or endothermic.

Many processes do not operate successfully over the range of temperatures that would be encountered in an adiabatic reactor, and in this case provision has to be made for addition or removal of heat. The most common way of doing this is to use a *tubular reactor*, with the catalyst in the tubes and a heat transfer medium surrounding them (see Fig. 1.4(b)). These are more expensive to make than adiabatic reactors, and changing the catalyst, when this becomes necessary, is much more time-consuming and expensive.

Another approach to heat transfer is to use a *fluidised-bed reactor* (see Fig. 1.4(c)). In this type of reactor, the reactants are passed upwards through a bed of catalyst at a velocity such that the catalyst particles are lifted by the gas stream, and the bed becomes fluidised. In this state it behaves in many respects like a liquid, and heat transfer can be carried out by coils or tubes immersed in the fluidised bed. This type of reactor is applied to only a limited extent in the petrochemical industry.

1.7.2 Product separation

In the petrochemical industry, the most important method of separation is distillation. Typically in continuous fractionation the mixture to be separated is heated and fed to a distillation column at a point

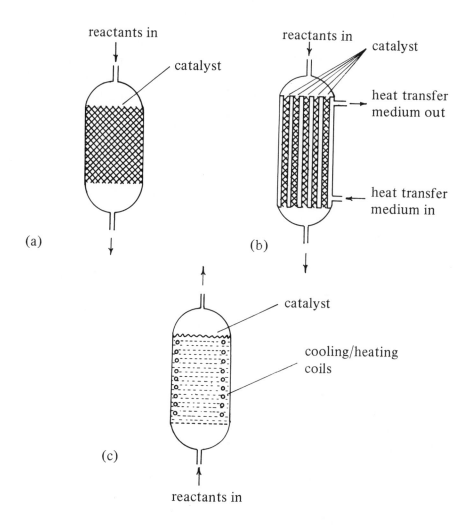

Fig. 1.4 — Catalytic reactors: (a) adiabatic reactor;
(b) tubular reactor; (c) fluidised-bed reactor.

roughly midway along its length. The column contains either packing or trays to promote vapour—liquid contact. Fractionation occurs along the length of the column, the more volatile components passing to the top, and the less volatile to the bottom. When more than two components are to be separated, usually one is taken off as either the top or bottom product and the remaining mixture is passed to another column, and so on until the desired number of components have been separated. Separation of complex mixtures consequently requires a large number of columns.

Vacuum distillation is used for materials which would decompose or undergo some other reaction at the temperatures that would be involved in distillation at atmospheric pressure. Gases are often separated by distillation under pressure and/or at low temperatures.

Various other methods of separation are also used. We will discuss these as appropriate when we meet them in individual processes.

1.8 COSTS IN PETROCHEMICAL PROCESSING

There are two major types of cost associated with making a chemical, the cost of building the plant, called the *capital cost*, and the cost of operating the plant, called the *production cost*.

The **capital cost** is made up of the cost of all the items of equipment that go into the construction of a plant – reactors, distillation columns, pumps, compressors, instruments, storage tanks, piping, and so on – together with the cost of their erection and of the associated buildings and facilities. Capital costs for modern plants to operate the processes discussed in this book would be in the region £10 million to £300 million.

The **production cost** is made up of the following components:

Raw materials. For most petrochemical processes this is the dominant production cost, often accounting for around 70% to 80% of the total.

Services. This is the cost of steam, electricity, gas, water, compressed air, and various other services which may be supplied to the plant. In the main the services are involved in the transfer of energy around the process. Service costs vary quite widely between processes.

Labour and supervision. This covers the wages and salaries and social costs of process operators and plant management. Typically this cost is low for petrochemical processes.

Maintenance. This is made up of engineers' and fitters' wages and salaries and social costs, and the costs of materials and equipment required for maintaining the plant.

Overheads. There are many costs incurred in running a factory and a company that are not associated directly with an individual plant. A proportion of the cost of these is allocated to each plant as 'overheads'.

Depreciation. This item allows for the fact that a plant has a finite lifetime, after which it will be scrapped, and so will have consumed the original capital cost.

The term 'operating cost' is used in this book to denote total production costs excluding raw material costs.

1.8.1 Chemistry and costs

Process chemistry interacts with capital cost in quite a complicated way, and it would not be feasible to discuss this completely here. However, what we can do is consider some of the main desirable process features as far as capital costs are concerned.

We have already seen that in a continuous process each separate reaction stage requires a reactor, and it is obvious that, other things being equal, the more reactors are needed, the more they will cost. Consequently, it is desirable that processes should involve the minimum number of reaction stages. One of the major features of the development of petrochemical processes over the years has been a move towards single-stage processes.

It is desirable that the minimum amount of separation of product from other materials be required, since the more separation is necessary, the larger and/or more numerous the distillation columns and other items of separation equipment will have to be. The ideal process would yield product of acceptable quality direct from the reactor. Such processes are very rare. Usually the exit stream from the reactor contains, in addition to product, a number of other materials, for example, unreacted raw materials, solvent, by-products, or catalysts.

The use of 'mild' conditions of temperature and pressure is desirable. The cheapest temperatures in terms of capital cost are in the approximate range 30 to 240°C, since in this range steam (under pressure) can be used for heating, and water can be used for cooling. As the temperature goes beyond 240°C the capital cost involved undergoes a number of stepwise increases. Below about 30°C, refrigerated cooling is needed, which is much more expensive in terms of capital cost than the use of cooling water. Operation at elevated pressures increases capital costs not only because of the increased cost of fabrication of vessels, pipes, distillation columns etc., but also because compressors are expensive items of equipment.†

The use of toxic, corrosive, and solid materials in processes tends to increase capital costs, and the use of such materials is minimised so far as possible.

---

† With gas phase processes, up to about 30 atm., the extra costs arising from the increased wall thickness are balanced by reductions in cost which are a consequence of the smaller volumes of vessels etc., required.

It has already been pointed out that as far as production costs are concerned raw material costs usually dominate the situation. The interaction of these with process chemistry is quite straight-forward — the cheaper the raw materials used and the less of them consumed, the lower the cost. However, one point that is worth emphasising is that in the large tonnage operations that are typical of the petrochemical industry, processes which convert raw materials into low value co-products are particularly undesirable. The manu-facture of propylene oxide by the chlorohydrin route typifies such processes:

$$CH_2{=}CHCH_3 \xrightarrow{Cl_2/H_2O} \underset{\underset{Cl\ \ \ OH}{|\ \ \ \ |}}{CH_2CHCH_3} \xrightarrow{Ca(OH)_2} \underset{\underset{O}{\backslash\!/}}{CH_2CHCH_3}$$

$$+ \qquad\qquad\qquad +$$

$$HCl \qquad\qquad\qquad CaCl_2$$

In addition to converting propylene to propylene oxide, this converts chlorine and calcium hydroxide to calcium chloride, which is essen-tially valueless.

Sometimes the process chemistry has important implications in terms of other production costs. For example, the use of high reaction temperatures with endothermic reactions, or the necessity for extensive separation operations, will tend to give rise to high service costs. We will discuss examples of such effects as they arise in dealing with processes.

1.8.2 Costs and scale of operation

Petrochemical processes are very susceptible to *economies of scale.* That is to say, up to a limit both production costs and capital cost per annual tonne decrease as the scale of operation is increased. This has two important effects. Firstly, it results in petrochemical manufacture being concentrated in a relatively small number of plants. In the UK, for example, the number of plants operating a given petrochemical process is commonly in the range 1 to 3. Secondly, for any given product there is a *minimum economic scale* of plant, determined partly by the nature of the technology, and partly by the scale on which other, competitive, companies are operating.

The maximum scale at which plants are built is limited by market considerations and by technological limits on the size of items of equipment.

1.9 NOMENCLATURE

During recent years there has been a move in the educational field, particularly at the school level, towards the exclusive use of systematic nomenclature for chemical compounds. However, in industry, and in society in general, systematic nomenclature is by no means always used, and this situation seems unlikely to change a great deal in the foreseeable future. It would be a reasonably safe bet that in the year 2000 we will still be calling $\{CH_2CHCl\}_n$ PVC rather than PCE.

The view that has been taken here is that when discussing the applications of chemistry in the petrochemical industry it is desirable to use the terminology that is used in the industry, and this is the practice that has been followed. In cases where both systematic and non-systematic names are commonly used, the former are usually chosen. For example, for the lower alkenes, the following are used: ethylene, propylene, 1-butene, 2-butene, and isobutene.

PROBLEMS AND EXERCISES

1. Write down the structures of the C_6 hydrocarbons you would expect to be present in crude oil.
2. Draw an outline flow scheme for an oil refinery which is operating both catalytic reforming and catalytic cracking. Show which products are used (i) for gasoline blending and (ii) as petrochemical feedstocks.
3. If the dehydrogenation stage of the styrene process discussed in section 1.7 was operated with no recycle of ethylbenzene, what would the yield of styrene (based on ethylbenzene) be? Do you think it likely that the process would be operated in this way?
4. In a process for the production of chlorobenzene, the feed rates of benzene and chlorine to the reactor are 5600 kg hr^{-1} and 3196 kg hr^{-1} respectively. The reactor effluent contains 4280 kg hr^{-1} monochlorobenzene and 2240 kg hr^{-1} benzene. Chlorine is totally consumed. The process stream passes to the separation section of the plant, where the monochlorobenzene and benzene are both recovered with 97% efficiency; the benzene is recycled. Draw an outline flow scheme for the process, and calculate (i) the conversion of benzene in the reaction stage, (ii) the yield of monochlorobenzene based on benzene, and (iii) the yield of monochlorobenzene based on chlorine.
5. Draw a diagram showing the arrangement of distillation columns that would be necessary to continuously separate a mixture of benzene, chlorobenzene, and o-dichlorobenzene.

2

Ethylene and co-products by thermal cracking

The manufacture of ethylene by thermal cracking is the prime example of a basic building block petrochemical process. It is carried out on a larger scale than any other organic chemical process. Total world capacity for ethylene is about 50 million tonnes, and USA and UK capacities are about 17.5 million tonnes per annum and 1.5 million tonnes per annum respectively. Modern crackers typically have capacities in the range 200 000 to 650 000 tonnes ethylene per annum.

2.1 FEEDSTOCKS

We have already noted that the main feedstocks used for the manufacture of ethylene are ethane, propane, naphtha, and gas oil; butane is also used to a minor extent. The relative importance of these materials varies markedly depending on location.

In areas where wet natural gas is available in ample quantities, ethane and, to a lesser extent, propane normally predominate, since they are technically very convenient raw materials, and are usually relatively cheap. The initial development of ethylene manufacture in the USA was very largely based on ethane and propane.

Being gases, ethane and propane are expensive to transport over long distances, and so in regions remote from sources of wet natural gas, an alternative raw material is necessary. Any of the fractions from crude oil distillation, up to and including gas oil, can be cracked to ethylene, though the technical problems increase somewhat with increase in molecular weight. During the period when the petrochemical industry began to develop outside the USA, naphtha was relatively cheap, since, in countries other than the USA, much more was being produced by refineries than was required for gasoline manufacture. Consequently, when thermal cracking was introduced in Western Europe and then in Japan, it was based on naphtha. This broad pattern, with ethylene manufacture being based on ethane and propane in the USA, and on naphtha elsewhere, persisted until around 1970. Since then, the situation has become more complicated.

One important change has been that under the influence of increasing demand for use for gasoline and as a petrochemical feedstock, the price of naphtha has risen relative to the other crude oil fractions. Since 1972, naphtha has, for most of the time, been more expensive than gas oil, though still somewhat cheaper than kerosine. This caused attention to be turned to the use of gas oil as a feed, and a limited number of crackers have been built that operate on gas oil or on naphtha or gas oil. The capital cost of a gas oil cracker is higher than that of a naphtha cracker, and the yield of ethylene from gas oil is lower than from naphtha. It appears that gas oil becomes an acceptable feed for ethylene manufacture when it is about 10% cheaper than naphtha.

In the USA, the availability of ethane and propane from natural gas has been declining since 1967, and consequently naphtha and, to a lesser extent, gas oil have come into increasing use as ethylene feedstocks. At the time of writing they account for about 45% of ethylene capacity in the USA. A change in the opposite direction is to some extent occurring in Western Europe, since ethane has become available from the associated gas from the North Sea oil fields, and ethylene is now being made from ethane in the UK. The proportion of ethylene derived from ethane in Western Europe is still, however, very small.

Finally, and potentially of very great significance to the petrochemical industry, a major new ethane-based ethylene industry is being set up in the Middle East. This will use ethane from associated gas that would otherwise be flared, i.e. burnt to waste.

2.2 THE REACTION SYSTEM

In essence, what is done in a thermal cracker is very simple: the feed is heated to a temperature in the range 750 to 900°C, cooled down again within less than 1 second, and the products are separated. However, achieving these reaction conditions poses some difficult engineering problems. Cracking technology is a specialised field, and there are only a limited number of companies that design ethylene plants.

The reaction is carried out in *cracking furnaces*. These are essentially tubular reactors as we discussed in section 1.7.1, with the tubes being heated direct by burning fuel. The detailed design of furnaces varies from company to company, and with feedstock to be processed, but in general terms, the tubes are from 5 to 20 cm in diameter, and 50 to 80 m in length. They are arranged in vertical folds in a refractory brick-lined furnace which is heated by burning gas or fuel oil. One of the main considerations in designing a cracking furnace is to extract the maximum possible amount of heat from the fuel used, and to this end it is arranged that the gas flow is preheated in tubes passing through the hot flue gases leaving the combustion zone, before they pass into the cracking tubes which are directly heated by the burning fuel (see Fig. 2.1). These two sections of the furnace are called the *convection section* and the *radiant section* respectively.

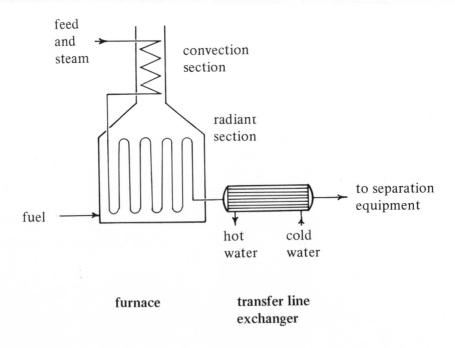

Fig. 2.1 — Cracking furnace.

The feed to the furnace is a mixture of hydrocarbon and steam, which acts as an inert diluent. The proportion of steam used increases with the molecular weight of the feed (see Table 2.1). The process stream is heated to about 600°C in the convection section, and then passes to the radiant section, where it spends between 0.15 and 0.75 seconds, and reaches a temperature of up to about 900°C. The rate of heat transfer needed to supply the sensible heat to raise the temperature and also the energy for the exothermic cracking reactions is high, and there is consequently a substantial temperature difference between the tube walls and the gas. Tube skin temperatures of up to 1100°C are involved, and special alloy steels with high chromium and nickel contents have to be used to stand up to the conditions.

Table 2.1
Typical steam ratios in cracking

Feed	kg steam/kg hydrocarbon
ethane	0.2–0.4
propane	0.3–0.5
naphtha	0.4–0.8
gas oil	0.8–1.0

When the product stream leaves the cracking furnace it is necessary for it to be cooled extremely rapidly to quench the reaction. At the same time, it is desirable to recover as much heat as possible to keep down energy costs. The precise way in which this is done depends on the feedstock being used, and on the particular plant design, but normally part of the cooling is done in a *transfer line exchanger* or *quench boiler*. This is a large heat exchanger, that is, a bundle of metal tubes through which the gases pass and around which is circulated water under pressure. The hot water produced is used to generate steam for use in the plant.

There is a limit to the size of cracking furnace which it is feasible to build, so that ethylene plants have a number of furnaces, all feeding into the same separation train. A 500 000 tonne per annum plant would commonly have in the region of ten furnaces.

Typical product distributions from ethane, propane, naphtha, and gas oil cracking are shown in Table 2.2.

Table 2.2
Typical product distributions in cracking (wt %)

	Feedstock			
	Ethane	Propane	Naphtha	Gas oil
H_2	3.6	1.3	0.8	0.6
CH_4	4.2	24.7	15.3	10.6
C_2H_2	0.2	0.3	0.7	0.4
C_2H_4	48.2	34.5	29.8	24.0
C_2H_6	40.0[†]	4.4	3.8	3.2
C_3H_4		0.3	1.1	1.0
C_3H_6	1.3	14.0	14.1	14.5
C_3H_8		10.0	0.3	0.4
$1,3\text{-}C_4H_6$			4.8	4.7
C_4H_8	1.6	3.7	4.2	4.5
C_4H_{10}			0.3	0.1
pyrolysis gasoline	0.9	5.9	21.0	18.4
fuel oil	—	0.9	3.8	17.6

† Unconverted ethane is recycled.

2.3 THE PRIMARY CRACKING REACTIONS

Thermal cracking involves free radical chain reactions. The steps involved in the initial formation of alkenes from the feedstock components, the so-called primary cracking reactions, are fairly well understood, and will be discussed in some detail in this section. The further reactions of the initially formed alkenes, called secondary reactions, are much less well characterised. They will be considered briefly in section 2.4.

2.3.1 Ethane cracking

At cracking temperatures, carbon—carbon bonds readily undergo homolysis, and initiation, i.e. the generation of free radicals, occurs as follows:

$$CH_3CH_3 \longrightarrow 2CH_3 \cdot$$

The alternative initiation reaction, involving carbon—hydrogen homolysis, is much slower, and is of negligible importance.

Abstraction of hydrogen atoms from ethane molecules by the methyl radicals gives rise to ethyl radicals:

$$CH_3\cdot + CH_3CH_3 \longrightarrow CH_4 + CH_3CH_2\cdot$$

In alkyl radicals, bonds β to the carbon atom bearing the unpaired electron are weaker than similar bonds in the corresponding alkane, and under cracking conditions such radicals readily undergo scission across these bonds to give a smaller radical and an alkene. This reaction, called β-scission, is the fundamental cracking reaction.

In an ethyl radical, β-scission can only occur across a carbon–hydrogen bond:

$$H-CH_2-CH_2\cdot \longrightarrow H\cdot + CH_2{=}CH_2 \qquad (1)$$

The hydrogen atom thus formed can now abstract a hydrogen atom from a futher molecule of ethane:

$$H\cdot + CH_3CH_3 \longrightarrow H_2 + CH_3CH_2\cdot \qquad (2)$$

Reactions (1) and (2) are the propagation steps in the kinetic chain. They occur a number of times for each initiation step.

A variety of termination steps can occur, e.g.

$$H\cdot + CH_3CH_2\cdot \longrightarrow CH_3CH_3$$

$$CH_3\cdot + CH_3CH_2\cdot \longrightarrow CH_3CH_2CH_3$$

$$CH_3CH_2\cdot + CH_3CH_2\cdot \longrightarrow CH_3CH_2CH_2CH_3$$

The products containing more than two carbon atoms that are formed in ethane cracking arise partly from termination reactions.

2.3.2 Propane cracking

In propane cracking, carbon–carbon homolysis gives rise to methyl and ethyl radicals:

$$CH_3CH_2CH_3 \longrightarrow CH_3\cdot + \cdot CH_2CH_3 \qquad (3)$$

The methyl radicals can, as we have seen, abstract hydrogen atoms from hydrogen molecules. So too can the ethyl radicals, but these can also undergo β-scission to give a hydrogen atom and a molecule of ethylene. Consequently, the hydrogen atom abstraction reactions that follow propane homolysis may be by methyl radicals, ethyl radicals, or hydrogen atoms:

$$C_3H_8 + CH_3 \cdot \text{ or } C_2H_5 \cdot \text{ or } H \cdot$$
$$\longrightarrow C_3H_7 \cdot + CH_4 \text{ or } C_2H_6 \text{ or } H_2$$

Since, as in ethane cracking, a number of propagation steps occur for each initiation step, most molecules of propane are attacked by a radical formed in a propagation step rather than one resulting from reaction (3). As we shall see, this means that attack will most frequently be by a hydrogen atom or a methyl radical.

Let us now consider the position of attack on the propane molecules. In terms of ease of abstraction, attack on the $-CH_2-$ hydrogens is favoured. On the other hand, there are three times as many CH_3- hydrogens, and the attacking species are relatively unselective. Consequently, both types of hydrogen are attacked, and both primary and secondary propyl radicals are formed:

$$C_3H_8 \xrightarrow{-[H \cdot]} \begin{array}{l} \nearrow CH_3CH_2CH_2 \cdot \\ \searrow CH_3\dot{C}HCH_3 \end{array}$$

The secondary propyl radicals can undergo carbon–hydrogen β-scission, to give propylene and hydrogen atoms:

$$CH_3\dot{C}HCH_3 \longrightarrow CH_3CH{=}CH_2 + H \cdot \qquad (4)$$

In the case of the primary radicals, there are two possibilities, carbon–hydrogen β-scission, and carbon–carbon β-scission, giving ethylene and a methyl radical:

$$CH_3 \overset{\frown}{-} CH_2 \overset{\frown}{CH_2} \cdot \longrightarrow CH_3 \cdot + CH_2{=}CH_2 \qquad (5)$$

Since carbon–carbon bonds are much weaker than carbon–hydrogen bonds, most primary propyl radicals undergo the latter reaction.

The hydrogen atoms and methyl radicals formed in reactions (4) and (5) then continue the chain by abstracting hydrogen atoms from further molecules of propane.

2.3.3 Naphtha and gas oil cracking

Most of the following discussion applies to both naphtha and gas oil, but to avoid tedious repetition of the phrase 'and gas oil', we shall refer to naphtha only.

Naphthas are, as we have seen, complex mixtures of alkanes, cycloalkanes, and aromatic hydrocarbons. It is convenient to talk about the chemistry of naphtha cracking in terms of the cracking of model compounds representing each of the main types of component.

Straight chain alkanes

Let us take nonane as a model for the straight chain alkane components. As is the case with ethane and propane, initiation involves homolysis of carbon–carbon bonds. For example:

$$C_9H_{20} \longrightarrow C_3H_7 \cdot + C_6H_{13} \cdot$$

The radicals thus formed can undergo carbon–carbon β-scission reactions:

$$CH_3CH_2CH_2 \cdot \longrightarrow CH_3 \cdot + CH_2{=}CH_2$$

$$CH_3CH_2CH_2CH_2CH_2CH_2 \cdot \longrightarrow CH_3CH_2CH_2CH_2 \cdot + CH_2{=}CH_2$$

Provided no other reaction interferes, these will continue until the end of the carbon chain is reached, to leave either a methyl radical or an ethyl radical:

$$CH_3CH_2CH_2CH_2 \cdot \longrightarrow CH_3CH_2 \cdot + CH_2{=}CH_2$$

The ethyl radicals can undergo carbon–hydrogen β-scission, though this is a much slower reaction than carbon–carbon β-scission.

The sequence of reactions that we have considered thus far may all be regarded as part of the initiation step of the cracking process. Most molecules of nonane do not start their reaction by undergoing homolysis, but rather by suffering hydrogen atom abstraction, most often by methyl or ethyl radicals or hydrogen atoms. This abstraction will usually generate a secondary radical, since the $-CH_2-$ hydrogens are most readily abstracted and are also most abundant:

$$C_9H_{20} \xrightarrow{\;-[H\cdot]\;} \text{e.g. } CH_3CH_2CH_2CH_2CH_2CH_2CH_2\dot{C}HCH_3$$

Carbon–carbon β-scission of this radical will give an alkene other than ethylene, and a smaller primary radical:

$$CH_3CH_2CH_2CH_2CH_2CH_2CH_2\dot{C}HCH_3$$
$$\longrightarrow CH_3CH_2CH_2CH_2CH_2CH_2 \cdot + CH_2{=}CHCH_3$$

This primary radical can then, as we have seen, undergo a series of β-scission reactions, each of which will give rise to a molecule of ethylene:

$$CH_3CH_2CH_2CH_2CH_2CH_2\cdot \longrightarrow H\cdot + 3CH_2{=}CH_2$$

The hydrogen atom or small radical formed at the end of this series of reactions can then continue the kinetic chain by abstracting a hydrogen atom from a further molecule of nonane.

It can be seen that the highest yield of ethylene will be obtained if, once formed, radicals undergo the maximum possible number of β-scission reactions. Consequently, is is desirable to minimise both termination reactions and *chain transfer reactions*, that is, reactions in which radicals abstract hydrogen atoms from nonane or other components of the reaction mix rather than undergo further β-scission.

Since both the termination and chain transfer reactions are bimolecular, their rates may be reduced relative to the rate of the unimolecular β-scission reactions by reducing the pressure at which cracking is carried out. For a number of reasons it is technically inconvenient to operate at reduced absolute pressure, and so steam is used as a diluent to reduce the partial pressure.

The other process variable which has a major effect on the yield of ethylene is cracking temperature. The β-scission reactions and the homolysis of alkane molecules have higher activation energies than the chain transfer and termination reactions, and consequently their rates relative to these reactions increase as the temperature is increased. This implies that the higher the reaction temperature, the higher the ethylene yield that is achievable. In fact, the maximum cracking temperature that can be used is determined by the maximum service temperature of the materials of construction of the cracking tubes, and by furnace design, and is currently about 900°C. This has been one of the major areas of development in naphtha cracking technology, and since 1950 the maximum temperatures achievable have increased by about 150°C, with a corresponding increase in ethylene yield from about 20 wt % to about 30 wt %.

Branched chain alkanes

From what we have seen, it can be appreciated that branched chain alkanes will give lower yields of ethylene than their straight chain isomers. Following through one reaction path for 4-ethylheptane will illustrate this:

$$CH_3CH_2CH_2\overset{\overset{\displaystyle CH_2CH_3}{|}}{CH}CH_2CH_2CH_3 \xrightarrow[\;]{-[H\cdot]} CH_3CH_2CH_2\overset{\overset{\displaystyle CH_2CH_3}{|}}{CH}CH_2\overset{\cdot}{C}HCH_3$$

$$CH_3CH_2CH_2\overset{\overset{\displaystyle CH_2CH_3}{|}}{CH}CH_2\overset{\cdot}{C}HCH_3 \longrightarrow CH_3CH_2CH_2\overset{\overset{\displaystyle CH_2CH_3}{|}}{C}H\cdot + CH_2{=}CHCH_3$$

$$CH_3CH_2CH_2\overset{\cdot}{C}HCH_2CH_3 \longrightarrow CH_3CH_2\cdot + CH_2{=}CHCH_2CH_3$$

$$CH_3CH_2\cdot \longrightarrow CH_2{=}CH_2 + H\cdot$$

Cycloalkanes

It is clear that cycloalkanes will also give lower yields of ethylene than the corresponding straight chain alkanes.

Cyclohexane and its homologues give conjugated dienes amongst the primary cracking products by reactions of the following type:

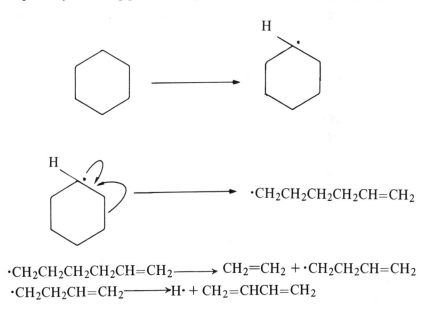

$$\cdot CH_2CH_2CH_2CH_2CH{=}CH_2 \longrightarrow CH_2{=}CH_2 + \cdot CH_2CH_2CH{=}CH_2$$
$$\cdot CH_2CH_2CH{=}CH_2 \longrightarrow H\cdot + CH_2{=}CHCH{=}CH_2$$

Aromatic hydrocarbons

The aromatic ring is essentially stable under cracking conditions, and in the main cracking occurs only in the side chains.

2.4 SECONDARY REACTIONS

The alkenes formed in the primary cracking reactions can undergo further secondary reactions. These are much less well understood than the primary reactions. Overall, they may be summarised as follows:

> lower alkenes
> dienes
> alkenes ———→ acetylenes
> aromatics
> coke

Thus, in cracking of propane and higher molecular weight feedstocks, secondary reactions result in both removal of ethylene, and its formation from other alkenes. Consequently, the proportion of ethylene and other alkenes produced varies with the *severity* of cracking, that is, the extent to which the overall reactions are allowed to proceed. Initially, as the severity is increased the yield of ethylene increases at the expense of the higher alkenes, but eventually the yield of ethylene also falls. Fig. 2.2 shows how product distribution varies with severity in naphtha cracking.

In ethane cracking, the secondary reactions do not produce any ethylene, so it would be desirable to completely eliminate them.

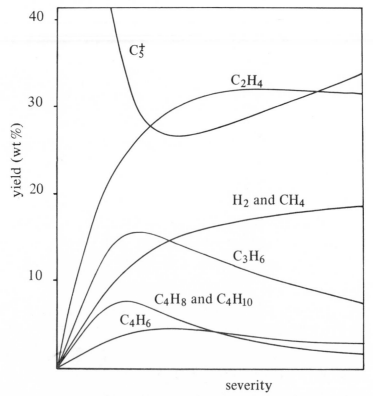

Fig. 2.2 – Naphtha cracking – variation
of product distribution with severity.

This could be achieved by operating at very low conversions of ethane, which would of course carry an economic penalty in separation and recycling costs. Conversions of 50% to 60% are commonly used, at which level ethylene is consumed in secondary reactions to only a fairly minor extent.

Whatever the feedstock, the formation of carbon is an undesirable outcome of secondary reactions, since it coats the walls of the cracking and transfer line exchanger tubes and impedes heat transfer and, ultimately, leads to blockages. In addition to improving ethylene yield, the use of steam diluent helps keep down the rate of carbon deposition, partly by bringing about the following reaction:

$$C + H_2O \longrightarrow CO + H_2$$

2.5 PRODUCTS AND PRODUCT SEPARATION

Crackers are built primarily for the purpose of producing ethylene. In the case of ethane crackers, this is essentially all they do produce, other than methane and hydrogen, which are usually used as fuel. The other feedstocks all give substantial amounts of co-products. In these cases, the product distribution, and in particular the ethylene—propylene ratio, can be varied, within limits, by carying the severity of cracking. Use is made of this in matching the output of a cracker to fluctuations in market requirements. With naphtha and gas oil crackers, the product distribution is also affected quite markedly by feedstock composition. Usually, a feed with a high alkane content, and particularly a high straight chain alkane content is preferred.

The cracker products are fractionally distilled to separate ethylene, and, depending on the feedstock, propylene, a C_4 stream, pyrolysis gasoline, and fuel oil. Elevated pressures and low temperatures are required for the ethylene and propylene separations.

The complexity of the separation section of a cracker increases markedly as the feed changes from ethane, through propane, to naphtha and gas oil, and this increased complexity is reflected in the capital costs of crackers (Table 2.3).

Table 2.3
Relative capital costs of crackers

Feedstock	Relative capital cost
Ethane	1.0
Propane	1.2
Naphtha	1.4
Gas oil	1.5

The return obtained for co-products has a major effect on the economics of propane, naphtha, and gas oil crackers. So far as possible, chemical outlets are found for all the propylene, though if the worst comes to the worst, it can be used as fuel. With naphtha and gas oil crackers, individual C_4 hydrocarbons are separated to varying extents for use in chemical manufacture (see Chapter 5). Pyrolysis gasoline can be used in motor gasoline, but is also a major source of aromatic hydrocarbons (see Chapter 6).

PROBLEMS AND EXERCISES

1. Use the data in Table 2.2 to calculate molar proportions of hydrogen, methane, and ethylene from ethane cracking. How would this product distribution change if conditions were such that the kinetic chain length of the reaction was reduced (i.e. that the number of propagation steps per initiation step went down)?

2. Assuming no losses in the separation section (which is of course unrealistic), what is the yield of ethylene, after ethane recycle, from an ethane cracker operating as indicated in Table 2.2?

3. Suggest mechanisms for the formation of (i) toluene, and (ii) styrene, in the thermal cracking of 1-phenylbutane.

4. What would you expect to be the major products formed in thermal cracking of (i) isobutane, and (ii) 2,2-dimethylpropane?

5. In catalytic cracking, the fundamental cracking reactions are very similar to those in thermal cracking, except that they involve carbocations rather than free radicals. From what you know of the chemistry of carbocations, suggest why the yield of ethylene from catalytic cracking is low.

6. Can you think of one of the reasons why cracking furnaces are not operated at reduced absolute pressure. (Hint: Think of safety.)

3

Ethylene derivatives

As we have already seen, petrochemical manufacture of ethylene began in the USA in the 1920s, and was largely confined to that country until after World War II. During this period, ethylene from non-petrochemical sources, e.g. from the dehydration of ethanol produced by fermentation, was to some extent used in other countries, notably in Germany, but its relatively high cost limited its applications. When, during the 1950s and the 1960s, petrochemical ethylene became widely available its use grew at a spectacular rate.

There have been two strands in the development of the applications of ethylene. One has been the growth of its use for making products, such as polyethylene, ethylene oxide, and styrene, which have been ethylene derivatives ever since they were commercially introduced. The other has been the development of ethylene-based processes for products, such as vinyl chloride, acetaldehyde, and vinyl acetate, which were originally made from other raw materials, notably from acetylene and from fermentation ethanol. Up to the mid-1960s these other routes were still of substantial importance. They are now largely obsolete, and ethylene is the supreme two-carbon building block.

We will discuss the major products made from ethylene roughly in order of their commercial importance.

3.1 POLYETHYLENE

The manufacture of polyethylene is the most important outlet for
ethylene, accounting for about 50% of consumption in the USA,
and rather more than that in Western Europe. Polyethylene is the
largest tonnage polymer made. US capacity is about 7.6 million
tonnes per annum.

There are three distinct types of polyethylene. The original
type, now called low density polyethylene, often abbreviated to
ldpe, was discovered by workers at ICI in the UK 1933, and was
first produced commercially in 1938. It is a relatively soft, flexible
plastic, probably most familiar as clear wrapping film. High density
polyethylene or hdpe, a harder, more rigid plastic, typically used
for articles like buckets and washing-up bowls, was introduced in
the mid-1950s. Linear low density polyethylene, which has similar
properties and uses to low density polyethylene, was introduced
towards the end of the 1970s. This group of products provides an
interesting example of the effects of process chemistry on polymer
structure, and hence on polymer properties and applications.

Low density polyethylene is made by free radical polymerisation.
The reaction is carried out at temperatures from 80 to 300°C and
pressures in the range 1000 to 3000 atm. Either oxygen or an organic
peroxide is used as the initiator. Initiation and propagation occur as
follows:

$$Initiator \longrightarrow 2R\cdot$$

$$R\cdot + CH_2{=}CH_2 \longrightarrow RCH_2CH_2\cdot$$

$$RCH_2CH_2\cdot + CH_2{=}CH_2 \longrightarrow RCH_2CH_2CH_2CH_2\cdot$$

$$\xrightarrow{\text{etc.}} R(CH_2CH_2)_nCH_2CH_2\cdot$$

Termination occurs by coupling and disproportionation of free
radicals:

$$\sim\!\!\sim\!\!\sim CH_2CH_2\cdot + \cdot CH_2CH_2 \sim\!\!\sim\!\!\sim$$
$$\longrightarrow \sim\!\!\sim\!\!\sim CH_2CH_2CH_2CH_2 \sim\!\!\sim\!\!\sim$$

$$\sim\!\!\sim\!\!\sim CH_2CH_2\cdot + \cdot CH_2CH_2 \sim\!\!\sim\!\!\sim$$
$$\longrightarrow \sim\!\!\sim\!\!\sim CH_2CH_3 + CH_2{=}CH \sim\!\!\sim\!\!\sim$$

If the above reactions were the only one to occur during the
polymerisation, the product would be linear polyethylene, with the

structure $+CH_2CH_2+_n$. In fact, low density polyethylene does not have this structure. Its molecules, far from being linear, are quite highly branched. Typically they have one or two long branches and a much larger number of short branches of up to five carbon atoms. These branches result from chain transfer reactions, in which hydrogen atom abstraction results in the active centre being transferred from one position to another. In intermolecular chain transfer, the free radical site is transferred from the growing radical to a finished polymer molecule. This leads to the formation of a long branch:

$$\sim\!\!\sim\!\!\sim CH_2CH_2\!\cdot\;+\;\sim\!\!\sim\!\!\sim CH_2CH_2\!\sim\!\!\sim\!\!\sim$$

$$\longrightarrow\;\sim\!\!\sim\!\!\sim CH_2CH_3\;+\;\sim\!\!\sim\!\!\sim\dot{C}HCH_2\!\sim\!\!\sim\!\!\sim$$

$$\longrightarrow\;\sim\!\!\sim\!\!\sim CHCH_2\!\sim\!\!\sim\!\!\sim$$
$$\underset{\displaystyle CH_2}{\overset{\displaystyle |}{}}$$
$$CH_2$$

In intramolecular chain transfer, the radical site is moved from one position on the radical to another a few carbon atoms away, resulting in the formation of a short branch:

$$\cdot CH_2\!-\!CH_2$$
$$\sim\!\!\sim\!\!\sim CH_2 \quad CH_2 \longrightarrow \sim\!\!\sim\!\!\sim \underset{C_4H_9}{CH\cdot} \longrightarrow \sim\!\!\sim\!\!\sim \underset{C_4H_9}{CH}\!\sim\!\!\sim\!\!\sim$$
$$CH_2$$

As we will see shortly, the branching has a profound effect on the properties of the polymer.

The pressures used in the manufacture of low density polyethylene are exceptionally high for a chemical process. Since the use of such pressures is expensive both in terms of capital costs and of operating cost, the question arises as to why they are used. The reason is that it is only at very high pressures, i.e. at high ethylene concentrations, that the propagation reactions are fast enough for high molecular weight polymer to be formed.

High density polyethylene is made by an entirely different type of process, called *co-ordination* or *Ziegler polymerisation*. This was discovered by Karl Ziegler in 1953, and the first commercial production of hdpe was about two years after that.

The catalyst systems developed by Ziegler consist of complexes formed from aluminium alkyls and transition metal halides, typically triethylaluminium and titanium tetrachloride. In the presence of such catalysts, polymerisation of ethylene occurs at about 100°C and at moderate pressures. Originally the reaction was carried out in an inert solvent such as hexane, but now it is also carried out in the gas phase. Quite separately from Ziegler, other workers developed systems for carrying out this type of polymerisation over transition metal oxide catalysts.

The mechanism of co-ordination polymeristation is complicated. We will simply consider it in outline.

The polymer chain grows at active metal sites on the catalyst, molecules of ethylene being first co-ordinated to the site, and then inserted between it and the growing chain:

$$M-CH_2CH_2\text{\Large\char"223B\char"223B\char"223B} + CH_2{=}CH_2 \longrightarrow M-\underset{\underset{CH_2{=}CH_2}{\big|}}{CH_2CH_2}\text{\Large\char"223B\char"223B\char"223B}$$

$$\longrightarrow M-CH_2CH_2CH_2CH_2\text{\Large\char"223B\char"223B\char"223B} \xrightarrow{\text{etc.}}$$

Chain transfer to polymer does not occur, and the polymer obtained is linear. It is this fact that accounts for the difference between its properties and those of low density polyethylene.

One of the reasons why polymers have useful mechanical properties is that there are large attractive forces between the very big polymer molecules. In non-polar polymers like polyethylene the attractions are due to van der Waal's forces. In other polymers, dipole–dipole interactions or hydrogen bonding may be involved. It can be appreciated that the more closely and regularly the polymer molecules can pack together, the stronger will be the forces between them. In high density polyethylene, with its regular linear structure, the molecules can pack together very closely in orderly arrays, called *crystalline regions*, for most of their lengths. In low density polyethylene the branches tend to get in the way, and a much smaller proportion of the total length of the molecules is in close packed ordered regions. This results in the polymer being softer, less strong, and having a lower melting point than the linear polymer, and also, as is indicated by the name, in its having a lower density.

The two products have different ranges of applications, reflecting their different properties and also the fact that low density polyethylene has traditionally been cheaper than high density polyethylene. This price relationship may seem strange in view of the costs associated with the operating pressures in the low density polyethylene process. It arose from the fact that, originally at least, the cost of making and handling the catalyst in the hdpe process, and the cost of removal of catalyst residues from the polymer, more than outweighed the cost of high pressure operation. With modern high activity catalysts this is probably no longer the case, and the price differential probably reflects the fact that for many applications high density polyethylene is the superior product.

The situation became more complicated in the late 1970s when the misleadingly named **linear low density polyethylene** (lldpe) was introduced. This is made by copolymerising ethylene with small amounts of 1-alkenes, such as 1-butene or 1-hexene, in the presence of Ziegler catalysts. It has short branches, introduced by the comonomer, but no long branches. By varying the nature and proportion of the comonomer, a substantial degree of 'tailoring' of polymer properties is possible. It is expected that linear low density polyethylene will to a considerable extent displace the conventional low density polymer from the market.

3.2 VINYL CHLORIDE

Vinyl chloride is required for the manufacture of poly (vinyl chloride), the second largest tonnage commercial polymer. About 14% and 18% of ethylene are used in the manufacture of vinyl chloride in the USA and Western Europe respectively.

Commercial manufacture of vinyl chloride was started in the early 1930s, using a process based on the reaction of acetylene with hydrogen chloride:

$$CH{\equiv}CH + HCl \xrightarrow[100-180°C]{HgCl_2 \text{ on charcoal}} CH_2{=}CHCl$$

This process has a number of attractive features. It has only one stage, uses quite mild conditions, does not need any supply of heat to the reaction stage since it is exothermic, and gives quite a high yield ($c.$ 90%). It does have the serious disadvantage, however, that acetylene, whether prepared from calcium carbide or by hydrocarbon cracking (section 10.6) is expensive. Ethylene is a much more attractive starting material.

A route from ethylene to vinyl chloride has been available since the mid-1930s:

$$CH_2{=}CH_2 + Cl_2 \xrightarrow{FeCl_3} CH_2ClCH_2Cl \xrightarrow{500°C} CH_2{=}CHCl + HCl$$

The first stage is a typical electrophilic addition of a halogen to an alkene. The second stage is a free radical chain reaction involving the following initiation and propagation steps:

$$CH_2ClCH_2Cl \longrightarrow Cl\cdot + \cdot CH_2CH_2Cl$$

$$Cl\cdot + CH_2ClCH_2Cl \longrightarrow HCl + \cdot CHClCH_2Cl$$

$$\cdot CHClCH_2Cl \longrightarrow CHCl{=}CH_2 + Cl\cdot$$

To avoid loss of yield, by further reaction of the vinyl chloride, the reaction is operated at 50% to 60% conversion, with recycle of unconverted dichloroethane. Yields of up to 99% are obtained.

This process has been used to a limited extent. However, there is a major problem associated with it: it converts half the chlorine fed to it to hydrogen chloride, for which there are only limited outlets. This greatly restricted its application. Acetylene continued to be the basis of most vinyl chloride manufacture well into the 1960s, although by this time it was very much more expensive than ethylene.

The situation was transformed by the development of **oxychlorination processes** for the production of dichloroethane. These involve the reaction of ethylene with hydrogen chloride and air or oxygen:

$$CH_2{=}CH_2 + 2HCl + \tfrac{1}{2}O_2 \xrightarrow[250-350°C]{CuCl_2/KCl} CH_2ClCH_2Cl + H_2O$$

Yields and conversions in oxychlorination are high, in the region of 90% and 95% respectively. The dichloroethane is converted to vinyl chloride by pyrolysis at about 500°C, and the hydrogen chloride produced in this stage is recycled to the oxychlorination reactor. Virtually all vinyl chloride manufacture is now by processes involving oxychlorination.

Overall, there are a number of ways in which processes of this type can be operated. In a situation where there is sufficient by-product hydrogen chloride available from some other process the plant could simply consist of an oxychorination stage and a pyrolysis stage. Where such a supply of hydrogen chloride is not available, or is available in insufficient quantity, it is usual to operate a balanced chlorination–oxychlorination process:

This is not the best conceivable way of making vinyl chloride from ethylene. The process has three stages and, as we saw in Chapter 1, two or one would be preferred. There has been much research activity in the field, and various other versions of oxychlorination processes have been developed. For example, one allows the use of mixtures of chlorine, hydrogen chloride and air for making dichloro-ethane, so that overall only two reactors are required. Another carries out the oxychlorination at about 450°C so that both oxy-chlorination and pyrolysis occur in the same reactor and vinyl chloride is obtained in one stage. However, at the time of writing, none of these processes appears to be operated commercially.

US capacity for vinyl chloride is about 4 million tonnes per annum.

3.2.1 Uses of vinyl chloride

By far the most important use of vinyl chloride is in the manu-facture of poly(vinyl chloride) and vinyl chloride copolymers. Poly(vinyl chloride) is made by free radical polymerisation:

$$CH_2{=}CHCl \longrightarrow \left[CH_2CH \atop \quad\ Cl \right]_n$$

It is a very versatile polymer since its properties can be varied over a wide range by the addition of plasticisers (section 8.4.1). Unplasti-cised poly(vinyl chloride) is a hard, rigid plastic. Plasticised poly(vinyl chloride) may be leathery or rubbery, depending on the proportion of plasticiser used.

The most important copolymer of vinyl chloride is vinyl chloride—vinyl acetate copolymer, which is mainly used for making gramophone records.

Some vinyl chloride is used for the manufacture of 1,1,1-trichloro-ethane (see section 3.8).

3.3 ETHYLENE OXIDE AND DERIVATIVES

Ethylene oxide is one of the oldest-established ethylene-based petro-chemicals. It has been made in the USA since about 1925. Its manu-facture accounts for about 17% and 12% of total ethylene consump-tion in the USA and Western Europe respectively.

The initial method of manufacture of ethylene oxide was the chlorohydrin process. This involves reacting ethylene with chlorine and water to give an aqueous solution of ethylene chlorohydrin and hydrogen chloride, which is then treated with calcium hydroxide to give ethylene oxide:

$$CH_2{=}CH_2 \xrightarrow[10-50°C]{Cl_2/H_2O} \begin{matrix} CH_2ClCH_2OH \\ + \\ HCl \end{matrix} \xrightarrow[100°C]{Ca(OH)_2} \underset{O}{CH_2CH_2} + CaCl_2$$

The yield of ethylene oxide obtained is about 85%.

This process has the very serious disadvantage of converting chlorine and calcium oxide into calcium chloride, for which there is very little demand. As we saw in Chapter 1, processes of this type which involve downgrading ancillary raw materials are avoided wherever possible in large-scale manufacture.

An alternative became available in the 1930s, when it was shown that ethylene could be directly oxidised to ethylene oxide by air or oxygen over a silver catalyst:

$$CH_2{=}CH_2 + O_2 \xrightarrow[\text{Ag catalyst}]{250-330°C} \underset{O}{CH_2CH_2}$$

Processes based on this reaction were first put into operation in the 1940s and progressively displaced the chlorohydrin route. For many years now, virtually all ethylene oxide has been made by direct oxidation.

The catalysts used are finely divided silver on a solid support such as alumina or silicon carbide. Ethylene oxide is removed from the gas stream leaving the reactor by scrubbing with water, and is recovered by distillation. Yields initially were around 60%, but continued process development has allowed them to be increased to up to 75%.

All gas phase oxidations are potentially hazardous in that under inadequately controlled conditions explosions may occur. Ethylene oxide manufacture presents unusually severe problems. One reason for this is that ethylene oxide is particularly susceptible to explosive reactions, even in the absence of oxygen. Great care has to be taken with heat removal and temperature control in the process. The reaction is carried out in tubular reactors typically containing several thousand tubes of 20 to 50 mm diameter.

US capacity for ethylene oxide is about 2.8 million tonnes per annum.

Ethylene oxide has a variety of uses as a chemical intermediate, all of which depend on the reactivity of the oxirane ring towards nucleophiles. We shall discuss the three most important uses.

3.3.1 Ethylene glycol

Ethylene glycol is made by reacting ethylene oxide with water:

$$\underset{\diagdown \! \diagup}{\underset{O}{CH_2CH_2}} + H_2O \longrightarrow CH_2OHCH_2OH$$

This is by far the most important application of ethylene oxide, accounting for about 60% of consumption.

In the absence of a catalyst, a temperature of about 200°C is required for an adequate rate of reaction, and to keep the reactants in the liquid phase a pressure of about 12 atm. has to be used. If a small amount of sulphuric acid (c. 0.5%) is added to the water, the reaction proceeds at an adequate rate at about 70°C, and no pressure is required. The choice between these two processes is not clear-cut, since the advantage that the acid-catalysed process has in not requiring the use of pressure is countered by the disadvantage that corrosion-resistant materials must be used for construction of the plant, and that the sulphuric acid must be removed from the reaction mix before isolation of the ethylene glycol. Both processes are used.

Ethylene oxide reacts with alcohols. There is consequently a tendency towards formation of di- and polyethylene glycols when ethylene glycol is being made:

$$HOCH_2CH_2OH + \underset{\diagdown \! \diagup}{\underset{O}{CH_2CH_2}} \longrightarrow HOCH_2CH_2OCH_2CH_2OH$$

$$\xrightarrow{etc.} HO[CH_2CH_2O]_nCH_2CH_2OH$$

These reactions can be suppressed by the use of a large excess of water (about 20 moles/mole ethylene oxide). However, the use of such an excess would result in very high separation costs. A molar water to ethylene oxide ratio of between 5 and 8 to 1 is usually used, and the formation of some polyglycols is accepted.

The ethylene glycol and polyglycols are separated by distillation. The yield of ethylene glycol is about 90%.

During the 1970s, a process was brought into operation for the direct manufacture of ethylene glycol from ethylene by the following route:

$$CH_2{=}CH_2 + CH_3CO_2H \xrightarrow[\text{160°C, 30 atm.}]{\text{O}_2,\ \text{Te \& Br salts}} \begin{array}{c} HOCH_2CH_2OAc \\ + \\ AcOCH_2CH_2OAc \end{array}$$

$$\xrightarrow[\text{120°C}]{\text{H}_2\text{O}} HOCH_2CH_2OH + CH_3CO_2H$$

Since the yield given by this process is about 95%, it was expected to be much more economically attractive than the route *via* ethylene oxide. However, the plant which was built to operate the process was shut down after eighteen months because of 'severe operational difficulties', and the future of the process seems uncertain.

Ethylene glycol has two major uses, as an antifreeze agent for use in internal combustion engines, and in the manufacture of poly(ethylene terephthalate) (see section 8.5.1).

3.3.2 Ethanolamines

Ethylene oxide reacts readily with aqueous ammonia:

$$\underset{\underset{O}{\diagdown\diagup}}{CH_2CH_2} + NH_3 \longrightarrow HOCH_2CH_2NH_2$$

The initial product, ethanolamine, has two nucleophilic groups, and both could react with further ethylene oxide. However, the $-NH_2$ group is much the stronger nucleophile, and under the conditions used it is only this that reacts:

$$HOCH_2CH_2NH_2 + \underset{\underset{O}{\diagdown\diagup}}{CH_2CH_2} \longrightarrow (HOCH_2CH_2)_2NH$$

The diethanolamine thus formed can react further to give triethanolamine:

$$(HOCH_2CH_2)_2NH + \underset{\underset{O}{\diagdown\diagup}}{CH_2CH_2} \longrightarrow (HOCH_2CH_2)_3N$$

The proportion of the products is controlled to suit market requirements by the ratio of ethylene oxide to ammonia used, and by the reaction temperature. Temperatures in the range 50 to 275°C and pressures of up to 100 atm. are used.

The ethanolamines have a variety of applications as involatile water soluble organic bases. They are, for example, used in aqueous solution for scrubbing weakly acidic components such as carbon dioxide or hydrogen sulphide from gas streams:

$$CO_2 + HOCH_2CH_2NH_2 + H_2O \longrightarrow HOCH_2CH_2NH_3^+ \ HCO_3^-$$

Since the salts formed are decomposed at moderately elevated temperatures, both the scrubbing solution and the acidic gas can be recovered:

$$HOCH_2CH_2NH_3^+ \ HCO_3^- \xrightarrow[c.\ 100^\circ C]{} HOCH_2CH_2NH_2 + H_2O + CO_2$$

3.3.3 Polyoxyethylene derivatives

The tendency, which we have already noted, of ethylene oxide to form polymeric products is utilised in a number of applications. For example, non-ionic surface active agents are made by reaction of ethylene oxide with alcohols, phenols, or amines, e.g.:

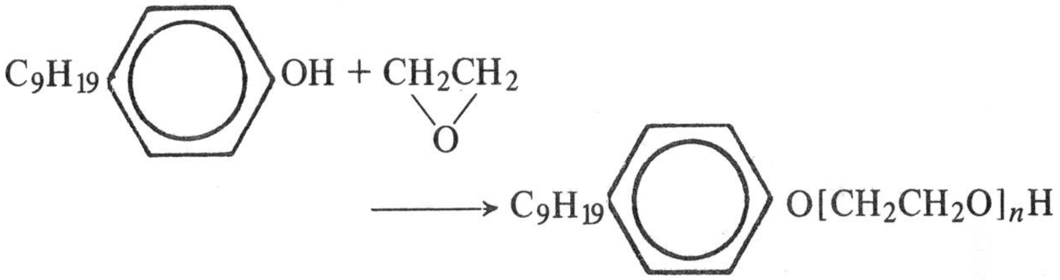

Polyoxyethylene glycols, water soluble polymers with a range of uses, are made by base catalysed reaction of ethylene oxide with water or ethylene glycol. Mixed polyoxyethylene–polyoxypropylene polyols are important in polyurethane manufacture.

3.4 STYRENE

The manufacture of styrene consumes about 6% of total ethylene in Western Europe, and about 7.3% in the USA. It is discussed in Chapter 7.

3.5 ETHANOL

Man has been making aqueous ethanol, for use as a beverage, for many thousands of years by fermentation of carbohydrates, and ethanol made in this way was one of the basic building blocks of the organic chemical industry until the 1950s. Now, however, most industrial ethanol is made by hydration of ethylene:

$$CH_2{=}CH_2 + H_2O \longrightarrow CH_3CH_2OH$$

In the USA, about 1.4% of ethylene is used in ethanol manufacture.

The original method of hydration, first used in the USA in 1930, was an indirect one, in which ethylene is absorbed in concentrated sulphuric acid to give a mixture of ethyl hydrogen sulphate and diethyl sulphate, which is then hydrolysed to ethanol:

$$CH_2=CH_2 \xrightarrow{\;H_2SO_4\;} \begin{array}{c} C_2H_5SO_4H \\ + \\ (C_2H_5)_2SO_4 \end{array} \xrightarrow{\;H_2O\;} C_2H_5OH + H_2SO_4$$

In the first stage, ethylene under a pressure of 20 to 35 atm. is passed into concentrated sulphuric acid at 60 to 90°C. (The purpose of the pressure is to increase the solubility of ethylene in the sulphuric acid, and thereby to increase the rate of reaction.) Water is then added, and the diluted mixture of sulphates is allowed to react at about 100°C. Ethanol is recovered and recycled. The yield of ethanol is about 90%, the main side reaction being the formation of diethyl ether.

This process has some rather obvious shortcomings. It has two reaction stages, it involves quite corrosive process streams, and the reconcentration of sulphuric acid consumes a substantial amount of energy. Direct catalytic reaction of ethylene with water would appear to be a much more elegant approach. A process of this type was first operated in 1947.

Direct hydration of ethylene requires the use of an acidic catalyst:

$$CH_2=CH_2 \xrightarrow{\;H^+\;} CH_3CH_2^+ \xrightarrow{\;H_2O\;} CH_3CH_2\overset{+}{O}H_2 \xrightarrow{\;-H^+\;} CH_3CH_2OH$$

The catalyst used is known as 'solid phosphoric acid'. It is essentially concentrated phosphoric acid held on an absorbent solid support. Ethylene and steam will react in contact with such a catalyst at an acceptable rate at about 300°C. However, there is a problem. The hydration reaction is reversible, and since it is exothermic the equilibrium constant decreases with increasing temperature. At 300°C, the equilibrium constant is very small. Consequently, if an equimolar mixture of steam and ethylene is allowed to react over the catalyst at this temperature and atmospheric pressure, the conversion of ethylene is very small.

Two methods of increasing the conversion suggest themselves: (i) the use of a large excess of water, and (ii) the use of elevated pressures. The first is not feasible, since at high partial pressures of water the phosphoric acid phase of the catalyst takes up water, and the catalyst loses its activity. The extent to which elevated pressures can be used is limited by the increased capital and operating costs involved, and by the fact that at high pressures formation of low polymers of ethylene becomes a troublesome side reaction. A third approach, the use of a catalyst which allows the reaction to be carried out at a lower temperature, is attractive, but no suitable catalyst has been found.

The process is typically operated at about 70 atm., with a molar ratio of water to ethylene of about 0.6:1. Under these conditions about 5% of ethylene is converted. The unconverted ethylene is recycled, and the overall yield of ethanol is about 95%. The main side reactions are polymerisation, mentioned above, and ether formation.

The low conversion obtained in the direct hydration process to some extent cancels out its advantages over the indirect process. Both processes are operated, though direct hydration is the favoured process for new plants. Total US capacity for synthetic ethanol is about 540 000 tonnes per annum.

Until the 1960s, the main use of ethanol was the manufacture of acetaldehyde, by dehydrogenation:

$$CH_3CH_2OH \longrightarrow CH_3CHO + H_2$$

For reasons which are discussed below, this is now of very minor importance. The main use of ethanol is now as a solvent, in a variety of applications. It is also used in the manufacture of a number of ethyl esters and other chemicals.

3.6 ACETALDEHYDE

Up to 1960 there were two commercial routes to acetaldehyde, dehydrogenation of ethanol, mentioned above, and the hydration of acetylene:

$$CH{\equiv}CH + H_2O \xrightarrow[70-90°C]{Hg^{2+}, \, dil.H_2SO_4} CH_3CHO$$

This latter route had been important, but by 1960 manufacture from ethanol accounted for a very large proportion of world production.

By 1970 this picture had totally changed, and most acetaldehyde was made directly from ethylene by a process called the **Wacker Chemie process**, after the name of the company where it was developed.

The story of how this development occurred is worth considering in a certain amount of detail.

During the mid-1950s, chemists at Wacker Chemie in Germany were investigating the oxidation of ethylene over palladium catalysts, with a view to developing an improved process for ethylene oxide. The system showed no promise for making ethylene oxide, but with some of the catalysts traces of acetaldehyde were formed. It was recognised that this could be important as a route to acetaldehyde, and the aim of the investigation was shifted.

The catalysts being used were palladium on charcoal, prepared by impregnating charcoal with palladium chloride and then reducing this to the metal. It was found that those catalysts which were leading to the production of acetaldehyde had been incompletely reduced, and still contained some palladium chloride. Further, it was found that as early as 1894 it had been reported that palladium chloride solutions oxidise ethylene to acetaldehyde:

$$PdCl_2 + CH_2{=}CH_2 + H_2O \longrightarrow Pd + 2HCl + CH_3CHO$$

Attempts to develop a gas phase catalytic oxidation did not give promising results, and attention was turned to a system using aqueous solutions. A process was developed in which a solution of palladium chloride and copper chloride in dilute aqueous hydrochloric acid is used as the catalyst. When ethylene is oxidised in such a solution, the following sequence of reactions occurs:

$$C_2H_4 + PdCl_2 + H_2O \longrightarrow CH_3CHO + Pd + 2HCl$$
$$Pd + 2CuCl_2 \longrightarrow PdCl_2 + 2CuCl$$
$$2CuCl + 2HCl + \tfrac{1}{2}O_2 \longrightarrow 2CuCl_2 + H_2O$$

The palladium chloride–ethylene reaction has been extensively investigated by the inventors of the process and others. It involves the formation of a π-complex between palladium chloride, ethylene and chloride ion followed by nucleophilic attack on this complex by water:

$$PdCl_2 + Cl^- + C_2H_4 \longrightarrow \left[\begin{array}{c} CH_2 \\ \| \\ CH_2 \end{array}\!\!\!-\, PdCl_3 \right]^-$$

$$\xrightarrow[\text{(multistep reaction)}]{H_2O} CH_3CHO + Pd + 2H^+ + 3Cl^-$$

Two versions of the process are operated. In the **one-stage process** (Fig. 3.1(a)), ethylene and oxygen, with the ethylene in substantial excess, are passed into the catalyst solution at about 125°C and under about 3 atm. pressure. The acetaldehyde (b.p. 21°C) is carried out of the reactor in the stream of ethylene and is recovered by scrubbing with water, followed by distillation. The excess ethylene is recycled. In the **two-stage process** (Fig. 3.1(b)) the catalyst solution is circulated round a system of two reactors. Ethylene at about

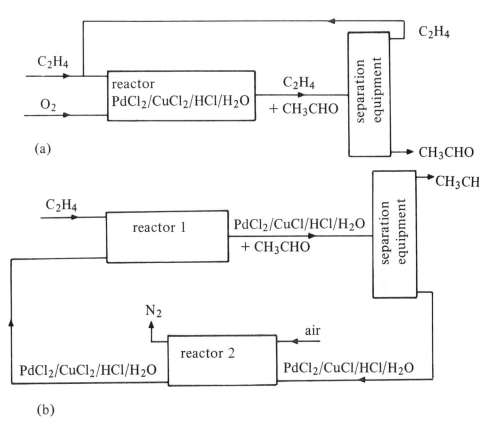

Fig. 3.1 — Wacker Chemie process: (a) one-stage; (b) two-stage.

10 atm. is passed into the first of these reactors and is oxidised by the catalyst solution. The system is operated so that the reduced catalyst solution emerging from this reactor contains palladium chloride and copper(I) chloride, but that metallic palladium is not formed. The catalyst solution, containing dissolved acetaldehyde, is then passed to a separator, where the pressure is reduced and the acetaldehyde boils off. The reduced catalyst solution then passes

to the second reactor, where it is reoxidised by air. Both versions of the process give yields of about 95%.

Since the two-stage process as described above is clearly more complicated than the one-stage process, and therefore would be expected to have higher capital and operating costs, one might ask why it should be used. To a large extent this is because the one-stage process requires oxygen, whereas the two-stage process uses air. The fair comparison is thus between the two-stage process and the one-stage process plus an oxygen plant. When such a comparison is made there is little to choose between the two.

That brings us to the question of why the one-stage process has to use oxygen.

For reasons of safety, it is necessary to operate the process with a gas mixture which is not explosive. This means either using a mixture containing an excess of ethylene, so that it is outside the higher explosive limit, or one containing a large excess of oxygen so that it is outside the lower explosive limit. Operation in this latter mode would involve two difficulties. Firstly, there would be a tendency for the acetaldehyde to be oxidised by the excess oxygen. Secondly, it would leave the reactor as a very dilute mixture with excess oxidising gas, and this would increase recovery costs. Consequently, as already noted, the process is operated with an excess of ethylene and the unreacted ethylene is recycled. If air was used as the oxidant, the excess ethylene would be mixed with nitrogen, and would have to be separated before it was recycled. This would involve either low temperature distillation or an extraction or adsorption process, all of which are relatively costly. It is economically preferable to use oxygen so that no separation of the ethylene is required before recycling.

The Wacker Chemie process shows very clear economic advantage over the alternative route from ethylene via ethanol, and has been extremely successful. Virtually all acetaldehyde is now made by this process. Rather more than 1% of ethylene goes to this application.

3.6.1 Uses of acetaldehyde

The main use of acetaldehyde is in the manufacture of acetic acid, by liquid phase air oxidation:

$$CH_3CHO \xrightarrow[\text{60--80°C, 5 atm.}]{\text{air, Co}^{2+}} CH_3CO_2H$$

The yield is about 95%. This is one of a number of important processes involving autoxidation or liquid phase free radical oxidation.

We shall discuss the mechanism of this type of reaction in sections 4.3 and 7.2.2. Oxidation of acetaldehyde in the presence of mixed cobalt and copper salt catalysts is also used, and produces a mixture of acetic acid and acetic anhydride.

There are other commercial routes to acetic acid, and the proportion made by the oxidation of acetaldehyde is diminishing. A process based on methanol, first operated in 1970, is the favoured process for new acetic acid plants (see section 9.6.2).

Acetaldehyde has a number of other applications as a chemical intermediate, but most of these use only relatively small tonnages. A fairly large-scale use, which has now, however, largely been replaced by another route (see section 4.5.3), was the manufacture of n-butyl alcohol via the aldol condensation:

$$CH_3CHO \xrightarrow{\text{base}} CH_3\overset{\overset{\displaystyle OH}{|}}{C}HCH_2CHO \xrightarrow{\text{acid}} CH_3CH{=}CHCHO$$

$$\xrightarrow{H_2} CH_3CH_2CH_2CH_2OH$$

3.7 VINYL ACETATE

Until the mid-1960s, vinyl acetate manufacture was based on acetylene:

$$CH{\equiv}CH + CH_3CO_2H \xrightarrow[\textit{c. } 200°C]{\text{Zn(OAc)}_2 \text{ on charcoal}} CH_2{=}CHO_2CCH_3$$

It is now made from ethylene, by a process which was developed as a logical extension of the Wacker Chemie acetaldehyde process. This involves oxidation of ethylene and acetic acid in the presence of a palladium catalyst:

$$CH_2{=}CH_2 + CH_3CO_2H + \tfrac{1}{2}O_2 \longrightarrow CH_2{=}CHO_2CCH_3 + H_2O$$

The mechanism is similar to that of the Wacker Chemie process, but with attack on the ethylene–palladium salt complex by acetate ions rather than by water.

The first version of the process, introduced in 1966, was closely similar to the acetaldehyde process. It involved operation in the liquid phase in acetic acid solution in the presence of palladium chloride, copper chloride, and sodium acetate. Three plants were built to operate this process, but all suffered from extremely severe corrosion problems, and all were abandoned after a short period.

The process which has been successfully operated, and which now accounts for most vinyl acetate manufacture, is a gas phase process in which a mixture of ethylene and acetic acid is oxidised over a palladium catalyst at around 200°C and 5 to 10 atm. A yield of about 95% is obtained.

US capacity for vinyl acetate is about 1.1 million tonnes per annum. Its manufacture consumes about 2.4% of ethylene.

Vinyl acetate is used in the manufacture of polymers and co-polymers. These have a very wide range of uses. For example, most household emulsion paints are based on copolymers of vinyl acetate and esters of acrylic acid.

3.8 CHLORINATED SOLVENTS

1,1,1-Trichloroethane, thrichloroethylene, and perchloroethylene are important solvents. They are all made from ethylene, though in the case of perchloroethylene there are other routes which are of major importance (see section 10.1 for one of these).

The main route to trichloroethane is based on vinyl chloride:

$$CH_2{=}CHCl \xrightarrow[\;40°C\;]{HCl,\,FeCl_3} CH_3CHCl_2 \xrightarrow[\;400°C\;]{Cl_2} CH_3CCl_3 + HCl$$

The yield is about 95%. As we have seen, virtually all vinyl chloride is made from ethylene.

There are a number of routes from ethylene to trichloroethylene and to perchloroethylene, some of which give both products. For example, chlorination of 1,2-dichloroethane at temperatures in the range 280 to 450°C gives a mixture of the two. However, it also gives large amounts of hydrogen chloride, which may be an embarrassment, and there is a growing tendency to use oxychlorination processes. Since the process situation is complicated, and manufacturers of these products are traditionally secretive about process technology, there is probably little further that can profitably be said.

The importance of these products arises from their good solvent properties coupled with lack of flammability and relatively low toxicity. Major uses are in metal degreasing, cleaning of electrical and electronic equipment, and in dry cleaning of clothing and household furnishings.

3.9 ETHYL CHLORIDE

Ethyl chloride is made from ethylene by reaction with hydrogen chloride, either in the gas phase at 130 to 250°C or in the liquid phase at about 35°C, in both cases in the presence of aluminium chloride:

$$CH_2=CH_2 + HCl \longrightarrow CH_3CH_2Cl$$

The yield is about 90%. In the USA in 1984, manufacture of ethyl chloride accounted for about 0.5% of ethylene consumption.

Some ethyl chloride is also made by chlorination of ethane:

$$CH_3CH_3 + Cl_2 \longrightarrow CH_3CH_2Cl + HCl$$

The economics of this process depend markedly on whether there is any outlet for the hydrogen chloride produced.

For many years the most important use of ethyl chloride has been in the manufacture of tetraethyl-lead:

$$4NaPb + 4C_2H_5Cl \longrightarrow Pb(C_2H_5)_4 + 4NaCl + 3Pb$$
(sodium lead
amalgam)

Tetraethyl-lead has only one significant application, as an antiknock additive for gasoline. The use of lead additives for this purpose has been severely reduced by legislation in recent years, and so the scale of this outlet has diminished greatly.

There are a number of other uses of ethyl chloride, mainly as a chemical intermediate, but they are all on a quite a small scale. We are not likely to see any ethyl chloride plants built during the next few years!

3.10 LINEAR 1-ALKENES AND LINEAR PRIMARY ALCOHOLS

The processes for these products are based on the so-called 'growth reaction' between ethylene and triethylaluminium. It was while he was studying this reaction that Ziegler discovered Ziegler polymerisation.

Ethylene reacts with triethylaluminium at moderate temperatures by successive insertions between the aluminium and the alkyl groups:

$$AlEt_3 \xrightarrow{C_2H_4} AlR_1R_2R_3 \qquad R = Et(CH_2CH_2)_n$$

Competing with this reaction is displacement of alkene by ethylene:

$$R_1R_2Al(CH_2CH_2)_nEt$$
$$\xrightarrow{C_2H_4} R_1R_2AlEt + CH_2=CH(CH_2CH_2)_{n-1}Et$$

The size of alkyl groups built up depends on the relative rates of the insertion and displacement reactions, and can be controlled by varying the conditions, notably the reaction temperature.

The linear alkene process is operated in two ways. In the two-step version, ethylene and triethylaluminium are reacted at about 100°C and 150 atm. Under these conditions the displacement reaction is slow, and the product is mainly a mixture of trialkylaluminiums. The temperature is then raised to about 250°C and the pressure reduced to about 15 atm., when the displacement reaction occurs, giving a mixture of alkenes and triethylaluminium, which is recycled.

In the one-step process, only a catalytic amount of triethyl-aluminium is used, and the reaction is carried out at about 250°C and 150 atm., under which conditions both growth and displacement occur. In a one-step process developed by Shell, a nickel complex is used as catalyst in place of triethylaluminium.

The alkenes are separated by distillation. Up to C_{12}, individual compounds are separated, but above that the products are mixtures, spanning ranges which depend on the intended applications. Typically products up to C_{18} are made.

Manufacture of linear alkenes by this route is carried out at only a fairly small number of locations. Plants tend to be large, with capacities in the range 175 000 to 200 000 tonnes per annum. The main use of the alkenes is in the production of alcohols by the Oxo process (section 4.5), and in the manufacture of detergent alkylate (section 7.6). Wax cracking (section 10.5) and dehydrogenation of linear alkanes (section 10.4) are alternative sources of linear alkenes.

Linear primary alcohols, often called 'fatty alcohols', may be made by oxidising the trialkylaluminiums produced by the growth reaction with dry air, to give a mixture of aluminium trialkoxides, which are then hydrolysed:

$$\underset{\diagdown R_3}{\overset{\diagup R_1}{Al-R_2}} \xrightarrow{O_2} \underset{\diagdown OR_3}{\overset{\diagup OR_1}{Al-OR_2}} \xrightarrow{H_2O} R_1OH + R_2OH + R_3OH + Al(OH)_3$$

In this case, the triethylaluminium is not recovered, and this adds substantially to costs. The process appears to be not very attractive compared to the alternative method of manufacture of fatty alcohols by hydrogenolysis of animal and vegetable fats and oils. It is operated to only a very limited extent.

PROBLEMS AND EXERCISES

1. Show the structure of one of the branches in a linear low density polyethylene made using 1-hexene as comonomer.

2. Draw an outline flow scheme for a balanced chlorination—oxychlorination vinyl chloride plant. Remember to show recycle streams.

3. Give mechanisms for the formation of ethyl hydrogen sulphate and diethyl sulphate from ethylene and sulphuric acid.

4. Suggest mechanisms for the formation of low polymers and diethyl ether in the direct hydration process for ethanol.

5. Calculate the raw material costs for acetaldehyde manufacture by (i) direct hydration of ethylene followed by dehydrogenation of ethanol and (ii) the Wacker Chemie process, assuming a yield of 90% in ethanol dehydrogenation and an ethylene cost of £320 per tonne.

6. The two-stage Wacker Chemie process can be operated on ethylene—ethane mixtures. Would it be feasible to operate the one-stage process in this way?

7. Give mechanisms for the first and second stages in the process for the manufacture of n-butyl alcohol from acetaldehyde.

8. In section 3.8 it was indicated that trichloroethylene and perchloroethylene may be made by chlorination of 1,2-dichloroethane. How would you expect the dichloroethane to be made?

4

Propylene derivatives

Whereas ethylene is produced in plants which were built specifically for that purpose, propylene is always obtained as a co-product, either of ethylene, from thermal crackers, or of gasoline, from catalytic crackers. Ethylene plants are the most important source in Western Europe and Japan, and catalytic crackers in the USA. There are two reasons for the difference. Firstly, in the USA, where there is a high demand for gasoline and a low demand for heavy fuel oil, catalytic cracking is carried out on a extremely large scale. Secondly, in Western Europe and Japan virtually all ethylene is produced by cracking naphtha or gas oil, whereas in the USA a substantial amount is made from ethane. As we saw in Chapter 2, ethane cracking does not produce any significant amount of propylene.

Normally, propylene supply exceeds demand, and this is reflected in its price, which has always been lower than that of ethylene. The total amount of propylene consumed in chemical applications is about 50% of ethylene consumption. Substantial amounts are also used in refinery processes for the manufacture of gasoline.

As is the case with ethylene, the use of propylene has developed along two lines. Some products, for example propylene oxide, have always been based on propylene. In other cases, for example that of acrylonitrile, propylene-based routes have displaced routes based on other intermediates.

The industrial chemistry of propylene is more diverse than that of ethylene since reactions both of the double bond and of the methyl group are involved.

4.1 POLYPROPYLENE

Polypropylene is the most recently introduced of the major plastics, having first been made commercially in 1956. Since then it has grown very rapidly in importance. Currently, it is the fourth largest tonnage polymer made. Its manufacture accounts for about 30% of total petrochemical propylene consumption.

The discovery of a method of making commercially useful polypropylene was made by Guilio Natta in Italy in 1954. Having heard of Ziegler's work on polyethylene, Natta started to investigate the polymerisation of propylene in the presence of Ziegler catalysts. He found that with certain catalyst systems high molecular weight polypropylene with useful mechanical properties was formed. This was in marked contrast to previous attempts to polymerise propylene, which had given either no high molecular weight polymer or polymer with no useful mechanical properties. Investigation of the structure of the polymer showed that it had a highly stereoregular structure, with all the methyl groups in the same steric configuration (Fig. 4.1). Natta coined the term 'isotactic polymer' for this type of material. Polymers of propylene made previously had no such stereoregularity, and in such polymers, called 'atactic' by Natta, the random orientation of the methyl groups totally prevents the close packing of polymer molecules into crystalline regions. Isotactic polypropylene on the other hand is highly crystalline. The importance of Natta's work was recognised by the award to him, jointly with Ziegler, of the Nobel prize for chemistry.

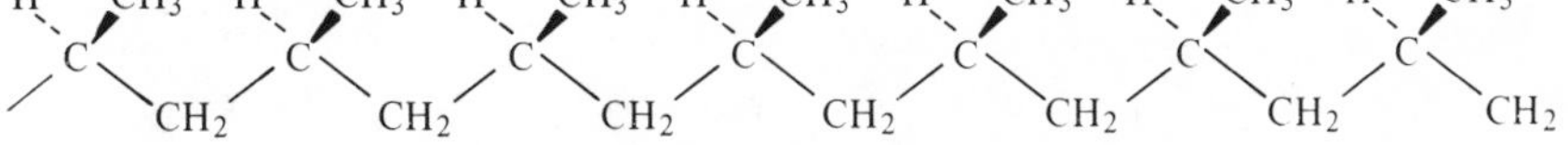

Fig. 4.1 – Isotactic polypropylene.

Polypropylene is made by processes similar to those used for high density polyethylene, and in fact some plants can make both polymers. It is used in both plastics and fibre applications.

US capacity for polypropylene is about 1.1 million tonnes per annum.

4.2 ISOPROPYL ALCOHOL AND ACETONE

As we have already seen, isopropyl alcohol has the distinction of having been the first petrochemical. Its manufacture currently consumes about 9% of propylene; about 50% of isopropyl alcohol is converted into acetone.

A large proportion of isopropyl alcohol is still made by essentially the same process that was first used in 1920, the sulphuric acid process. This is closely analogous to the process for ethanol which we have already discussed:

$$CH_3CH{=}CH_2 \xrightarrow[\text{45--55°C, 20--30 atm.}]{H_2SO_4} \begin{array}{c} i\text{-}PrSO_4H \\ + \\ (i\text{-}Pr)_2SO_4 \end{array}$$

$$\xrightarrow{H_2O} CH_3\underset{\underset{OH}{|}}{C}HCH_3$$

Since propylene is more reactive towards electrophiles than ethylene, milder reaction conditions can be used in the sulphation stage than in the ethanol process. In particular, the concentration of the sulphuric acid used is in the range 70% to 85% compared with around 96% in the ethanol process. The acid reconcentration costs are consequently significantly lower than those in ethanol manufacture.

The process is often operated on mixtures of propylene and propane, thus saving separation costs. In this case, the propane simply acts as an inert gas.

Direct hydration of propylene has been used since 1951:

$$CH_3CH{=}CH_2 + H_2O \longrightarrow CH_3\underset{\underset{OH}{|}}{C}HCH_3$$

There are problems associated with the position of the equilibrium in this reaction similar to those in ethylene hydration which we discussed in section 3.5. However, whereas the only catalyst used in ethanol manufacture is solid phosphoric acid, propylene can be hydrated over a variety of catalysts, and a number of processes are operated. Hydration over solid phosphoric acid is carried out at 200 to 250°C and about 40 atm., and gives a conversion of about 4%. A process developed by ICI uses a tungsten oxide catalyst, and operates at 250 to 300°C and 200 to 300 atm., with a molar water to propylene ratio of about 2.5:1. One developed by Deutsche Texaco uses an acidic ion exchange resin catalyst, and operates at 130 to 160°C and 80 to 100 atm. with a water to propylene ratio of about $14:1$. In these two processes, water is present in the liquid phase, and since the isopropyl alcohol dissolves in this as it is formed, the position of the equilibrium is displaced. Much higher conversions are consequently obtainable than in the purely gas phase solid phosphoric acid catalyst process. A conversion of 60% to 70% is claimed for the Deutsche Texaco process. Yields in all three processes are around 95%.

Acetone is produced by catalytic dehydrogenation of isopropyl alcohol over a copper–zinc catalyst at 400 to 500°C or over a zinc oxide catalyst at about 380°C:

$$CH_3\underset{\underset{OH}{|}}{C}HCH_3 \longrightarrow CH_3\underset{\underset{O}{\|}}{C}CH_3 + H_2$$

The process is operated at 70% to 85% conversion, when yields of over 90% are obtained after recycle. Higher conversions are possible, but then the yield is reduced by further reaction of the acetone.

Acetone can be made direct from propylene by oxidation in the presence of a palladium chloride–copper chloride catalyst solution (cf. acetaldehyde, section 3.6). This process has not achieved any major significance, in distinct contrast to the situation with acetaldehyde. The main reason for this is that for many years there has been no need to build plants specifically to make acetone, since large amounts of co-product material have been available. The cumene process for phenol is the most important source of co-product acetone (section 7.3.2), but some is also produced in the manufacture of acetic acid by oxidation of butane and naphtha (section 10.3). Co-product acetone accounts for more than 50% of production.

In 1983, US production of acetone from all sources was about 830 000 tonnes.

Apart from its use in making acetone, the main outlet for isopropyl alcohol is as a solvent.

Acetone is also used as a solvent, and as an intermediate in the manufacture of a large number of chemicals. The two most important of these are methyl methacrylate and methyl isobutyl ketone, made as indicated below:

$$CH_3COCH_3 \xrightarrow[25°C]{HCN, NaOH} CH_3\underset{\underset{OH}{|}}{\overset{\overset{CH_3}{|}}{C}}CN$$

$$\xrightarrow[100°C]{H_2SO_4} CH_2=\overset{\overset{CH_3}{|}}{C}CONH_2 . H_2SO_4 \qquad \xrightarrow[90°C]{MeOH} CH_2=\overset{\overset{CH_3}{|}}{C}CO_2CH_3$$

methyl methacrylate

$$CH_3COCH_3 \xrightarrow[0-20^\circ C]{base} CH_3COCH_2\overset{\overset{\displaystyle CH_3}{|}}{\underset{\underset{\displaystyle OH}{|}}{C}}CH_3$$

$$\xrightarrow[100^\circ C]{acid} CH_3COCH=\overset{\overset{\displaystyle CH_3}{|}}{C}CH_3 \xrightarrow[120-165^\circ C]{H_2, Ni \text{ or } Cu} CH_3COCH_2CH(CH_3)_2$$

methyl isobutyl ketone

Methyl methacrylate, US capacity about 540 000 tonnes per annum, is the monomer for the plastic poly(methyl methacrylate), sold under various tradenames, e.g. 'Perspex' and 'Plexiglas'. Methyl isobutyl ketone, US capacity about 390 000 tonnes per annum, is used as a solvent, mainly in surface coatings.

4.3 PROPYLENE OXIDE

Propylene oxide became of importance during the 1950s. Currently, its manufacture consumes about 10% of petrochemical propylene.

At first sight, by analogy with ethylene oxide, it might be expected that propylene oxide would be made by direct catalytic oxidation of propylene. In fact, all attempts to bring such an oxidation about in an acceptable yield have failed. The problem arises from the reactivity of the allylic hydrogens of the methyl group, which results in oxidation tending to occur at this position rather than at the double bond.

Until 1968, all propylene oxide was made by the chlorohydrin process, a direct analogue of the process initially used for ethylene oxide manufacture:

$$CH_3CH=CH_2 \xrightarrow{Cl_2/H_2O} \begin{matrix} CH_3\overset{\overset{\displaystyle |}{}}{C}HCH_2Cl \\ \overset{\displaystyle |}{OH} \\ + \\ HCl \end{matrix}$$

$$\xrightarrow{Ca(OH)_2} CH_3\overset{}{C}HCH_2 + CaCl_2$$
$$\underset{O}{\diagdown \diagup}$$

The yield is about 90%. As we saw in section 3.3, the fact that the chlorohydrin process converts large amounts of chlorine and calcium hydroxide into low-value calcium chloride is a major economic drawback.

Since 1968 an alternative process has been available, and has been coming into increasing use. The key step in this process, called the **Halcon process** after the American company which invented it, is reaction of propylene with a hydroperoxide:

$$CH_3CH=CH_2 + ROOH \longrightarrow \underset{\underset{O}{\diagdown\diagup}}{CH_3CHCH_2} + ROH$$

Two versions of the process are operated, one using t-butyl hydroperoxide and the other ethylbenzene hydroperoxide.

Hydroperoxides play a key role in many processes involving liquid phase free radical oxidation. Let us discuss the mechanism of their formation, using the oxidation of isobutane as an example.

Liquid phase free radical oxidation is a chain reaction. Reaction of a particular molecule of substrate starts with hydrogen atom abstraction by a free radical. For the moment, let us call this radical $X\cdot$. In the case of isobutane the tertiary hydrogen atom is much more readily abstracted than the methyl hydrogens, and under the reaction conditions normally used, this is the only position of attack:

$$(CH_3)_3CH + X\cdot \longrightarrow (CH_3)_3C\cdot + XH \tag{1}$$

Oxygen, a biradical, readily reacts with radicals:

$$(CH_3)_3C\cdot + O_2 \longrightarrow (CH_3)_3COO\cdot \tag{2}$$

The peroxy radical thus formed then abstracts a hydrogen atom from a further molecule of isobutane:

$$(CH_3)_3COO\cdot + (CH_3)_3CH \longrightarrow (CH_3)_3COOH + (CH_3)_3C\cdot \tag{3}$$

It can be seen that reactions (2) and (3) are propagation steps, and that if $X\cdot$ is a peroxy radical, reactions (1) and (3) are equivalent.

Once the reaction is under way, that is, once an appreciable concentration of hydroperoxide has built up, free radicals are generated by homolysis of the weak oxygen—oxygen bond of the hydroperoxide:

$$(CH_3)_3COOH \longrightarrow (CH_3)_3CO\cdot + \cdot OH \tag{4}$$

From this it can be seen that $X\cdot$ can also be $(CH_3)_3CO\cdot$ and $\cdot OH$, and that t-butyl alcohol would be expected to be one of the reaction

products. When the object is to prepare the hydroperoxide, conditions have to be adjusted so that reaction (4) occurs only to the
extent necessary to ensure a rate of generation of free radicals which
will lead to an acceptable rate of reaction. In other processes, decomposition of the hydroperoxide is a required feature of the reaction,
and is encouraged by use of an appropriate reaction temperature,
and often by the use of a metal salt catalyst (see section 7.2.2).

The question remains as to what is the source of free radicals
right at the start of the reaction, before any hydroperoxide has been
formed. One suggestion is that oxygen takes part in hydrogen atom
abstraction:

$$(CH_3)_3CH + O_2 \longrightarrow (CH_3)_3C\cdot + HO_2\cdot$$

The question is of no practical importance in manufacturing processes,
since the oxidations are carried out in continuous stirred tank flow
reactors, and hydroperoxide is always present.

In the Halcon process, isobutane or ethylbenzene are oxidised
with air at 120 to 150°C and about 30 atm. The reaction is taken to
about 25% conversion, and the yield of hydroperoxide is about 60%.
The hydroperoxide is then reacted with propylene in the liquid phase
at 120 to 140°C and around 35 atm., with a molybdenum salt
catalyst in solution. Thus, overall the two versions of the process
operate as follows:

$(CH_3)_3CH \longrightarrow (CH_3)_3COOH$ ⟶ $(CH_3)_3COH$
$+$
$CH_2=CHCH_3$ ⟶ CH_2CHCH_3
$\diagdown O \diagup$

$PhCH_2CH_3 \longrightarrow Ph\overset{\displaystyle OOH}{\overset{\displaystyle |}{C}}HCH_3$ ⟶ $Ph\overset{\displaystyle OH}{\overset{\displaystyle |}{C}}HCH_3$
$+$
$CH_2=CHCH_3$ ⟶ CH_2CHCH_3
$\diagdown O \diagup$

The yield of propylene oxide based on propylene is about 90%.

It can be seen from the above discussion that the alcohol ROH
is formed in both the oxidation stage and the epoxidation stage.
The amounts of alcohol produced are consequently large, about 3 kg

and 3.5 kg of *t*-butyl alcohol and 1-phenylethanol respectively per kilogram of propylene oxide. It is clearly essential to the economic viability of the process that there is a commercial outlet for these co-products. *t*-Butyl alcohol is used as an octane number improver for gasoline, and can also be dehydrated to isobutene. 1-Phenyl-ethanol is converted to styrene, by dehydration over an acidic catalyst, typically silica gel or titanium dioxide, at temperatures in the range 180 to 400°C:

$$
\underset{\overset{|}{\text{PhCHCH}_3}}{\overset{\text{OH}}{}} \longrightarrow \text{PhCH}=\text{CH}_2 + \text{H}_2\text{O}
$$

About 2.8 kg styrene is produced for every kilogram propylene oxide.

A number of large plants have been built to operate the Halcon process, and it now makes a very substantial proportion of the total production of propylene oxide. Shell also operate a process based on propylene oxide and ethylbenzene which differs from the Halcon process in using a solid catalyst in the epoxidation stage rather than one in solution. This is also operated on a large scale. It appears highly likely that these processes will totally displace the chloro-hydrin process.

US capacity for propylene oxide (all processes) is about 1.8 million tonnes per annum.

4.3.1 Uses of propylene oxide

Propylene oxide has two main uses, in making polyether polyols and propylene glycol, accounting for about 58% and 24% of consumption respectively.

Polyether polyols are made by reacting propylene oxide with polyhydroxy compounds in the presence of basic catalysts. The most important are based on reaction with glycerol:

$$
\begin{array}{l}
\text{CH}_2\text{OH} \\
| \\
\text{CHOH} + \text{CH}_2\text{CHCH}_3 \xrightarrow[\text{100--120°C}]{\text{KOH}} \\
| \qquad\qquad\quad \backslash\!/ \\
\text{CH}_2\text{OH} \qquad\quad \text{O}
\end{array}
\quad
\begin{array}{l}
\text{CH}_2[\text{OCH}_2\text{CH(CH}_3)]_x\text{OH} \\
| \\
\text{CH}[\text{OCH}_2\text{CH(CH}_3)]_y\text{OH} \\
| \\
\text{CH}_2[\text{OCH}_2\text{CH(CH}_3)]_z\text{OH}
\end{array}
$$

The polyols, with molecular weights typically in the range 3000 to 5000, are used in the manufacture of polyurethane foams.

Propylene glycol is made by reacting propylene oxide with water under conditions similar to those used for hydrating ethylene oxide (section 3.3.1). As in the case of ethylene glycol manufacture, poly-glycol formation is controlled at acceptable levels by using an excess of water. The main use of propylene glycol is in making unsaturated polyester resins.

4.4 CUMENE

The manufacture of cumene, for use in the production of phenol and acetone, consumes about 8% of petrochemical propylene. It is discussed in Chapter 7.

4.5 HYDROFORMYLATION: BUTYL ALCOHOL AND OTHER PRODUCTS

Hydroformylation involves the reaction of carbon monoxide and hydrogen with alkenes, to give aldehydes, e.g.

$$CH_3CH=CH_2 + CO + H_2 \longrightarrow CH_3CH_2CH_2CHO + CH_3\overset{\overset{\displaystyle CHO}{|}}{C}HCH_3$$

The reaction was discovered in Germany in 1938, and a plant was built there during World War II, but was never operated. After the war, details of the process were acquired by technical investigation teams from Britain and the USA, and plants based on this information were built in these countries.

The process, often called the *Oxo process*, is now of great importance in the petrochemical industry. Although the hydroformylation reaction itself produces aldehydes, the process is, in the main, operated for the purpose of making alcohols. A number of alkenes are used as feed, but the most important single one is propylene.

4.5.1 Conventional Oxo process

In the version of the process originally developed in Germany, the reaction is carried out in the liquid phase at 110 to 180°C and 200 to 300 atm. in the presence of a soluble cobalt catalyst. The cobalt is added to the reactor as an oxide or a salt, and the active catalytic species, which is believed to be either hydridotetracarbonylcobalt, $HCo(CO)_4$, or hydridotricarbonylcobalt, $HCo(CO)_3$, is formed in the reactor. These compounds are stable only under a high partial pressure of carbon monoxide and hydrogen; this is the reason for the use of high pressures in the process. The catalyst passes out of the reactor with the reaction product, and has to be recovered and recycled. Various methods are used for this. For example, metallic cobalt may be recovered by reducing the pressure so that the catalyst complex decomposes.

The mechanism of hydroformylation is complicated, and beyond the scope of this book.[†] The overall effect is the addition of H— and —CHO across the double bond. This addition occurs in both directions, but the straight chain aldehyde is the major product. Thus, with propylene, n-butyraldehyde and isobutyraldehyde are obtained in yields of around 67% and 15% respectively.

The aldehydes are separated by distillation, and are hydrogenated in either the gas or the liquid phase to the corresponding alcohols, e.g.

$$n\text{-}C_3H_7CHO + (CH_3)_2CHCHO$$
$$\xrightarrow{H_2} n\text{-}C_4H_9OH + (CH_3)_2CHCH_2OH$$

Various catalysts, e.g. nickel or cobalt, are used for the hydrogenation.

4.5.2 Newer processes

The pressures used in the conventional process are high, and this carries economic penalties in terms of both capital and operating costs. Another problem is that although the proportion of branched chain products is relatively small, in many cases it is still more than is desired, since the demand for branched chain alcohols is much more limited than that for straight chain alcohols. Processes which are improvements in both these respects have been developed.

In the Shell process, developed during the 1960s, a trialkyl-phosphine, e.g. tributylphosphine, is added to the reactor, and the active catalytic species is of the type $HCo(CO)_3PBu_3$. The phosphine ligand stabilises the complex, and consequently lower pressures can be used than in the conventional process. Higher ratios of straight chain products are obtained, and alcohols can be produced in one step, since the catalyst has hydrogenation activity. On the other hand, the catalyst is less active, so that higher temperatures have to be used. Also, some hydrogenation of the alkene occurs, which tends to nullify the effect of the higher proportion of straight chain products.

With a propylene feed, the process is operated at 160 to 200°C and 50 to 100 atm., and gives a yield of about 67% n-butyl alcohol and 8% isobutyl alcohol.

A more recently developed process, first brought into operation in 1975, uses a rhodium catalyst stabilised by triphenylphosphine.

[†] Discussions of the mechanism of hydroformylation may be found in the books by Gates *et al.*, Pearce and Patterson, and Sheldon in the 'Sources of further information, and further reading' list.

This requires a pressure of only around 20 atm. and a temperature of about 100°C, and gives a very high ratio of straight to branched chain products. Hydrogenation of the aldehydes is carried out in a separate stage, as in the conventional process. The main problem about the process is that since rhodium is a precious metal virtually no catalyst losses can be tolerated.

Conventional hydroformylation is still much the most important process.

4.5.3 Hydroformylation products

Apart from hydrogenation to n-butyl alcohol, the main use of n-butyraldehyde is the manufacture of 2-ethylhexanol:

$$C_3H_7CHO \xrightarrow{\text{base}} C_3H_7\underset{\underset{OH}{|}}{CH}\overset{\overset{Et}{|}}{CH}CHO \xrightarrow{\text{acid}} C_3H_7CH{=}\overset{\overset{Et}{|}}{C}CHO$$

$$\xrightarrow{\text{H}_2,\text{ catalyst}} C_3H_7CH_2\overset{\overset{Et}{|}}{CH}CH_2OH$$

2-Ethylhexanol is an important plasticiser alcohol (section 8.4.1).

Outlets for isobutyraldehyde are much more limited. The most important use is in the manufacture of isobutyl alcohol, and there are a number of other, small tonnage applications.

n-Butyl alcohol has two main uses, as a solvent in surface coatings, and for the manufacture of butyl esters, notably the plasticiser dibutyl phthalate. Isobutyl alcohol is also used as a solvent, though its properties are in general somewhat inferior to those of the straight chain alcohol. Isobutyl esters are also used to some extent as plasticisers.

The other major hydroformylation products are alcohols in the range C_7 to C_{19}, for use in making plasticisers and detergents.

4.6 PROPYLENE TRIMER AND TETRAMER

In the presence of acidic catalysts propylene yields low polymers. This reaction is used for the manufacture of mixed C_9 and C_{12} alkenes, called propylene trimer and propylene tetramer respectively.

The process is carried out in the vapour phase over a solid phosphoric acid catalyst at around 200 to 240°C and 15 to 25 atm. It involves the steps shown below:

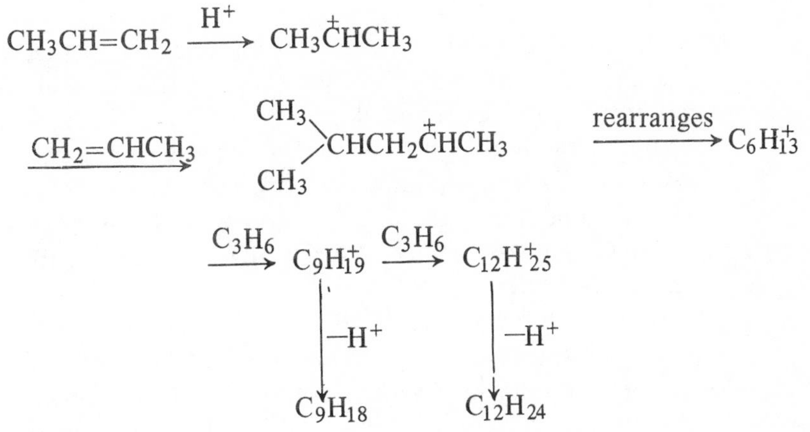

$$CH_3CH=CH_2 \xrightarrow{\ H^+\ } CH_3\overset{+}{C}HCH_3$$

$$\xrightarrow{CH_2=CHCH_3} \quad \underset{CH_3}{\overset{CH_3}{>}}CHCH_2\overset{+}{C}HCH_3 \quad \xrightarrow{\text{rearranges}} C_6H_{13}^+$$

$$\xrightarrow{C_3H_6} C_9H_{19}^+ \xrightarrow{C_3H_6} C_{12}H_{25}^+$$

$$\Big\downarrow -H^+ \qquad\qquad \Big\downarrow -H^+$$

$$C_9H_{18} \qquad\qquad C_{12}H_{24}$$

Since the carbocation intermediates rearrange rapidly, mixtures of highly branched alkenes are produced.

Formerly, the most important outlet for propylene trimer and tetramer was in the manufacture of surface active agents. For reasons discussed in section 7.6 this is now of very much less importance. The major use now is as feedstock for the Oxo process for the production of plasticiser alcohols.

4.7 ACRYLIC ACID AND ACRYLATES

Although esters of acrylic acid were used commercially in Germany before World War II, they remained of only minor importance until the 1950s. Since then they have become widely used in a variety of copolymers.

Initially, acrylic acid and its esters were made by a number of routes, all of which were based on relatively expensive raw materials. For example, one process was based on acetylene and carbon monoxide:

$$C_2H_2 + CO + H_2O \xrightarrow[100°C]{NiCl_2} CH_2=CHCO_2H$$

In 1969, the first plant to produce acrylic acid by the gas phase catalytic oxidation of propylene was put into operation, and this is now the main method of manufacture. The oxidation may be carried out in one or two stages. In the two-stage process, propylene is oxidised to acrolein, and then in a separate reactor this is oxidised further to acrylic acid:

$$CH_2=CHCH_3 \longrightarrow CH_2=CHCHO \longrightarrow CH_2=CHCO_2H$$

In the one-stage process, both reactions occur in the same reactor. The advantage of using two stages is that the catalyst and reaction conditions can be designed to be optimum for each of the two separate reactions, whereas in the one-stage process a compromise has to be accepted, and yields and conversions tend to be lower. At the same time, the capital cost will clearly be higher for the two-stage process, so as is often the case, an economic optimum has to be sought.

At present the two-stage process appears to be the favoured version. Typically, propylene is oxidised at about 320°C and the acrolein—air mixture is then fed to the second reactor, where the acrolein is oxidised at about 280°C. Multicomponent metal oxide catalysts are used in both stages. Acrylic acid is stripped from the gas stream by scrubbing with water, extracted from the water solution by a solvent such as butyl acetate, and then isolated by vacuum distillation. The yield of acrylic acid is in the range 73% to 83%.

Although some acrylic acid is used as such, as a comonomer for introducing carboxyl groups into polymers, most of it is converted into esters, e.g.

$$CH_2=CHCO_2H + C_2H_5OH \xrightarrow[c.\ 150°C]{} CH_2=CHCO_2C_2H_5$$

ethyl acrylate

Acrylates are used in the manufacture of a very wide variety of copolymers. As we have already noted, one very important application is in copolymers with vinyl acetate for use in emulsion paints.

US capacity for acrylic acid and acrylates is about 660 000 tonnes per annum.

4.8 ACRYLONITRILE

Acrylonitrile was first used commercially in Germany during the 1930s, for making nitrile rubber (section 5.3.1). This was only a moderate tonnage application, and acrylonitrile did not become of major importance until the introduction of acrylic fibres in 1950.

Until 1959, two processes were used for the manufacture of acrylonitrile, one based on acetylene and hydrogen cyanide, and the other, much less important, based on ethylene oxide and hydrogen cyanide:

$$C_2H_2 + HCN \longrightarrow CH_2=CHCN$$

$$\underset{\underset{O}{\diagdown\diagup}}{CH_2CH_2} + HCN \longrightarrow HOCH_2CH_2CN \longrightarrow CH_2=CHCN$$

In 1959, an entirely novel process based on the oxidation of propylene in the presence of ammonia was introduced:

$$CH_2=CHCH_3 + NH_3 + 1\tfrac{1}{2}O_2 \longrightarrow CH_2=CHCN + 3H_2O$$

This type of reaction was subsequently given the name 'ammoxidation'. Although the yield from the ammoxidation process was less than 60% when it was first introduced, the low cost of propylene and ammonia relative to acetylene, ethylene oxide, and hydrogen cyanide gave it an overwhelming economic advantage, with raw material costs of only about one-third those of the acetylene-based process. After its introduction the price of acrylonitrile fell dramatically, and operation of the older processes became uneconomic. Within a few years, virtually all world manufacture of acrylonitrile was by ammoxidation. Currently about 16% of petrochemical propylene is used in this application.

The invention of the ammoxidation process arose out of research on the oxidation of propylene to acrolein (as also did the acrylic acid process already discussed). Processes for oxidising propylene to acrolein were available by the beginning of the 1950s, and the potential availability of low cost acrolein stimulated consideration of its use for making acrylonitrile. The line of thought was as follows. Ammonia will react with the carbonyl group of acrolein to give, initially, an aldehyde ammonia:

$$CH_2=CHCHO + NH_3 \longrightarrow CH_2=CHCHNH_2 \atop \overset{\displaystyle OH}{|}$$

This has the correct carbon—nitrogen skeleton for acrylonitrile, and can be envisaged as being converted into it by a combination of dehydration and dehydrogenation:

$$\underset{\overset{|}{\underset{\displaystyle OH}{}}}{CH_2=CHCHNH_2} \xrightarrow{-H_2O,\ -H_2} CH_2=CHCN$$

It was found that acrylonitrile could in fact be made by reacting acrolein, ammonia, and oxygen over an appropriate catalyst. This opened up the possibility of a two-stage process from propylene to acrylonitrile. However, the companies investigating the route, the Distillers Company in the UK and Sohio in the USA, then went one step further and investigated the reaction of propylene, ammonia, and oxygen. Both developed ammoxidation processes. The Sohio process was commercialised first, and has dominated the technology ever since.

The original Sohio catalyst was bismuth phosphomolybdate, but now a complex, multicomponent catalyst containing bismuth and molybdenum together with a number of other elements is used. The reaction is carried out at 400 to 450°C in fluidised-bed reactors. Conversion of propylene is high, and the yield is over 70%. The main by-products are hydrogen cyanide, acetonitrile, and acrolein. Acrylonitrile is scrubbed out of the product gases with water, and separated by distillation.

US capacity for acrylonitrile is about 1 million tonnes per annum.

The main use of acrylonitrile is in the manufacture of acrylic fibres, e.g. Acrilan, Courtelle, Orlon. Acrylic fibres contain about 90% of acrylonitrile, together with other monomers to influence dyeing characteristics. Acrylonitrile is also used in nitrile rubber, ABS resins, and styrene–acrylonitrile copolymer.

A use which has developed fairly recently is in the manufacture of adiponitrile, one of the intermediates in nylon 66 production, by electrolytic hydrodimerisation:

$$CH_2{=}CHCN + 2H^+ + 2e^- \longrightarrow NC(CH_2)_4CN$$

The main route to adiponitrile is from adipic acid (section 7.2.4).

4.9 ALLYL CHLORIDE

Allyl chloride was first made commercially in 1945. Its manufacture involves the so-called 'hot chlorination' reaction discovered in Shell laboratories in the USA just before World War II:

$$CH_2{=}CHCH_3 + Cl_2 \xrightarrow[c.\ 500°C]{} CH_2{=}CHCH_2Cl + HCl$$

A large excess of propylene is used (about 4 moles C_3H_6 per mole Cl_2), and unreacted propylene is recycled. The yield of allyl chloride is about 85%.

The substitution occurs by a free radical chain reaction:

$$Cl_2 \longrightarrow 2Cl\cdot$$

$$CH_2{=}CHCH_3 + Cl\cdot \longrightarrow CH_2{=}CHCH_2\cdot + HCl \tag{5}$$

$$CH_2{=}CHCH_2\cdot + Cl_2 \longrightarrow CH_2{=}CHCH_2Cl + Cl\cdot \tag{6}$$

At 500°C, reactions (5) and (6) are much faster than the following reactions which result in addition:

$$Cl\cdot + CH_2=CHCH_3 \longrightarrow ClCH_2\dot{C}HCH_3$$

$$ClCH_2\dot{C}HCH_3 + Cl_2 \longrightarrow ClCH_2\overset{\overset{\displaystyle Cl}{|}}{C}HCH_3 + Cl\cdot$$

Allyl chloride is made on a much more limited scale than the other products discussed in this chapter. Its main uses are in the manufacture of epichlorohydrin and allyl alcohol:

$$CH_2=CHCH_2Cl \xrightarrow{Cl_2/H_2O} \begin{array}{c} HOCH_2CHClCH_2Cl \\ + \\ ClCH_2CH(OH)CH_2Cl \end{array}$$

$$\xrightarrow{NaOH} \underset{\displaystyle epichlorohydrin}{CH_2CHCH_2Cl \atop \overset{\backslash \ /}{O}}$$

Epichlorohydrin is used in the manufacture of epoxy resins and glycerol. Allyl alcohol has a number of small tonnage polymer applications.

PROBLEMS AND EXERCISES

1. Approximately how much propylene would be available from a naphtha cracker making 500 000 tonnes per annum of ethylene?
2. Draw a diagram illustrating the molecular structure of atactic polypropylene.
3. What by-product do you think might be formed in process for the direct hydration of propylene?
4. When ethylbenzene is oxidised to produce ethylbenzene hydro-peroxide, attack is predominantly on the $-CH_2-$ rather than the CH_3- group. Suggest why this is so.
5. In the above oxidation, acetophenone ($PhCOCH_3$) is obtained as a by-product. Write down a mechanism for its formation.
6. Suggest a mechanism for the dehydration of 1-phenylethanol.
7. Suggest a route for the manufacture of 1-propanol on a sub-stantial scale (e.g. about 20 000 tonnes per annum).

5

Butadiene and butenes

5.1 INTRODUCTION

The situation with regard to sources, separation, and uses of butadiene and butenes is much less clear-cut than that for either ethylene or propylene.

As we saw in Chapter 2, substantial quantities of butadiene and butenes are available in the C_4 stream from naphtha and gas oil crackers. They cannot, however, be separated simply by fractional distillation since the boiling points of the C_4 stream components are too close together (Table 5.1). Further, since outlets for the compounds, particularly for the butenes, are limited, it is by no means always the case that their isolation is required. Consequently, the extent to which individual components are separated from cracker C_4 streams, and the way in which it is done, vary considerably from plant to plant. Material not required for chemical applications is generally used as fuel.

Table 5.1
Boiling points of C_4 stream components

Hydrocarbon	b.p. ($°C$)
isobutane	−11.7
isobutene	−6.9
1-butene	−6.3
butadiene	−4.4
n-butane	−0.5
trans-2-butene	0.9
cis-2-butene	3.7

Butenes are also available in refineries, in catalytic cracker gases. Again, the extent to which they are separated for chemical use varies. Catalytic cracker C_4 gases not required for chemical manufacture are used in gasoline production, either by direct blending into gasoline, or via the refinery processes of alkyation or polymerisation, or are used in liquefied petroleum gas, or as refinery fuel.

5.2 BUTADIENE – MANUFACTURE

Most large naphtha and gas oil crackers incorporate a butadiene separation plant. Where other C_4s are to be isolated, the butadiene is normally separated first.

Methods of separation of butadiene depend on interactions between the conjugated π-electron system of the butadiene molecule and various polar solvents. The original method used solvent extraction by ammonaical copper(I) acetate solution, but modern processes all involve *extractive distillation*. In this, the C_4 mix is distilled in the presence of a solvent which decreases the volatility of the butadiene relative to the butenes and butanes. The solvent and butadiene pass from the bottom of the distillation column, and are then separated by distillation. Solvents commonly used are furfural, acetonitrile, *N*-methylpyrrolidone, and dimethylformamide.

In regions where the main feedstock for ethylene production is naphtha or gas oil, e.g. Western Europe, co-product butadiene tends to be available in excess over demand.

In the USA, the amount of co-product material is not sufficient to meet demand, and substantial quantities of butadiene are made by dehydrogenating n-butenes and n-butane:

$$n\text{-}C_4H_8 \longrightarrow CH_2{=}CHCH{=}CH_2 + H_2$$

$$n\text{-}C_4H_{10} \longrightarrow CH_2{=}CHCH{=}CH_2 + 2H_2$$

The dehydrogenations are carried out over metal oxide catalysts at around 650°C and at reduced pressure or with steam dilution. Yields and conversions vary considerably from process to process, and with the feed. Yields of 80% at conversions of about 35%, and 60% at conversions of about 30% may be taken as reasonably typical for butene and butane feeds respectively.

Dehydrogenation processes are becoming less important as a source of butadiene in the USA as the amount of co-product material available from naphtha and gas oil crackers increases.

5.3 BUTADIENE – USES

The major use of butadiene, accounting for well over 80% of consumption, is as a monomer, mainly for the manufacture of synthetic rubbers. It has a limited number of other uses, of which only two, the manufacture of chloroprene and of adiponitrile, are carried out on a substantial scale. Recently, BASF in Germany have developed a process for the manufacture of adipic acid from butadiene and carbon monoxide:

$$CH_2{=}CHCH{=}CH_2 + CO + MeOH \longrightarrow MeO_2C(CH_2)_4CO_2Me$$

$$\xrightarrow{H_2O} HO_2C(CH_2)_4CO_2H + MeOH$$

However, this has not yet been operated at the commercial scale, and it remains to be seen how important it becomes.

5.3.1 Synthetic rubbers

Styrene–butadiene rubber, or **SBR**, is the largest-tonnage synthetic rubber made. It was developed in Germany in the 1930s by IG

Farbenindustrie. Details of its manufacture were made available to du Pont in the USA under a know-how exchange agreement, and ironically it was on the basis of this information that the huge synthetic rubber industry that was built up in the USA during World War II was based.

SBR is a random copolymer of butadiene and styrene, normally in the ratio 76:24. It is mainly made by emulsion polymerisation, either at around 50°C using, for example, potassium persulphate as the initiator, or at around 5°C using a 'redox' system such as p-menthyl hydroperoxide and iron(II) sulphate. Solid polymer is isolated by coagulating the emulsion by the addition of acid or aluminium sulphate.

SBR is a *general-purpose rubber.* Its major single use is in making tyres. In 1983, US production of SBR was 904 000 tonnes.

Nitrile rubber, a copolymer of butadiene and acrylonitrile, was developed at about the same time as SBR, and is made by similar processes. It is one of the *speciality rubbers*, used in applications where its resistance to oil and chemicals is of value. Its scale of production is much smaller than that of SBR.

Polybutadiene rubbers are made by Ziegler–Natta polymerisation, and are stereoregular polymers containing a high proportion of cis-1,4 monomer units in the polymer chain. They were introduced in 1960, and have grown rapidly in importance, as general-purpose rubbers. US production in 1983 was 333 000 tonnes.

5.3.2 Chloroprene

Chloroprene is the monomer for neoprene, one of the earliest synthetic rubbers, first produced in 1932. The original method of manufacture was based on acetylene:

$$CH{\equiv}CH \xrightarrow[20°C]{CuCl\ aq.} CH{\equiv}CCH{=}CH_2$$

$$\xrightarrow[50°C]{conc.\ HCl/CuCl} CH_2{=}CClCH{=}CH_2$$

This has now largely been replaced by manufacture from butadiene. Chlorination of butadiene in the gas phase gives a mixture of 1,4-dichloro-2-butenes and 3,4-dichloro-1-butene. The 1,4-dichlorobutenes are isomerised to the 3,4-adduct, and the 3,4-dichloro-1-butene is dehydrochlorinated by treatment with aqueous sodium hydroxide:

$$CH_2=CHCH=CH_2$$

$$\xrightarrow[300°C]{Cl_2} CH_2ClCH=CHCH_2Cl + CH_2=CHCHClCH_2Cl$$

$$CH_2ClCH=CHCH_2Cl \xrightarrow[100°C]{CuCl} CH_2=CHCHClCH_2Cl$$

$$CH_2=CHCHClCH_2Cl \xrightarrow[100°C]{15\% NaOH} CH_2=CHCCl=CH_2$$

This process clearly lacks somewhat in elegance. It is, however, economically attractive compared with the acetylene-based process which suffers from the dual disadvantage of being based on an expensive hydrocarbon and of handling vinylacetylene, which can undergo explosive decompositions.

Neoprene is made by emulsion polymerisation of chloroprene, typically at about 40°C, using potassium persulphate as initiator. It is a speciality rubber, used in applications where its resistance to chemical attack and to solvents is important. US production of neoprene in 1983 was 115 000 tonnes.

5.3.3 Adiponitrile

We have already seen one route to adiponitrile, based on acrylonitrile, and we will discuss the main method of manufacture, from adipic acid, in section 7.2.4.

Manufacture from butadiene was started by du Pont in 1959. The original route involved chlorination of butadiene, followed by treatment of the dichlorobutenes obtained with sodium cyanide, to give 1,4-dicyano-2-butene, which was then hydrogenated. In 1971, du Pont introduced a process which avoids the use of chlorine. Published details of this process are scanty, but it appears to follow the overall route:

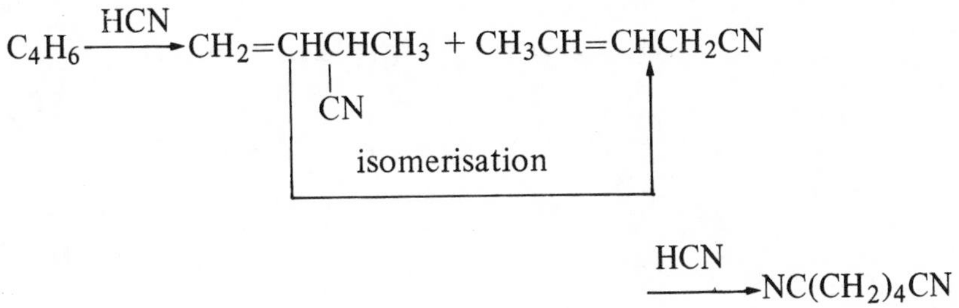

$$C_4H_6 \xrightarrow{\text{HCN}} \underset{\underset{\text{isomerisation}}{\overset{|}{CN}}}{CH_2=CHCHCH_3} + CH_3CH=CHCH_2CN$$

$$\xrightarrow{\text{HCN}} NC(CH_2)_4CN$$

All stages are carried out in the liquid phase at around 100°C in the presence of soluble nickel arylphosphite complexes.

5.4 ISOBUTENE

Isobutene is obtained from the C_4 stream from catalytic crackers, or from the butane–butene mix remaining after the removal of butadiene from naphtha cracker C_4 streams. Separation is achieved by taking advantage of the fact that isobutene is substantially more reactive towards electrophiles than the other butenes. Typically the gas stream is scrubbed with 45% to 65% sulphuric acid at temperatures in the range 20 to 50°C. The isobutene is absorbed and hydrated to *t*-butyl alcohol, while the butanes and *n*-butenes pass through. The *t*-butyl alcohol may be isolated by vacuum distillation and dehydrated over an acidic catalyst, e.g. alumina at about 370°C, or alternatively the isobutene may be regenerated by diluting the acid solution and raising the temperature.

As we saw in section 4.3, the Halcon process for propylene oxide now also provides a source of isobutene.

Until recently, the only major chemical use of isobutene was in making polymers, of which the most important is **butyl rubber**. This is a copolymer of isobutene with a small amount of isoprene (2-methylbutadiene), made by cationic polymerisation at about −90°C in the presence of aluminium chloride or boron trifluoride:

$$(CH_3)_2C=CH_2 + AlCl_3 + HCl \longrightarrow (CH_3)_3C^+ \ AlCl_4^-$$

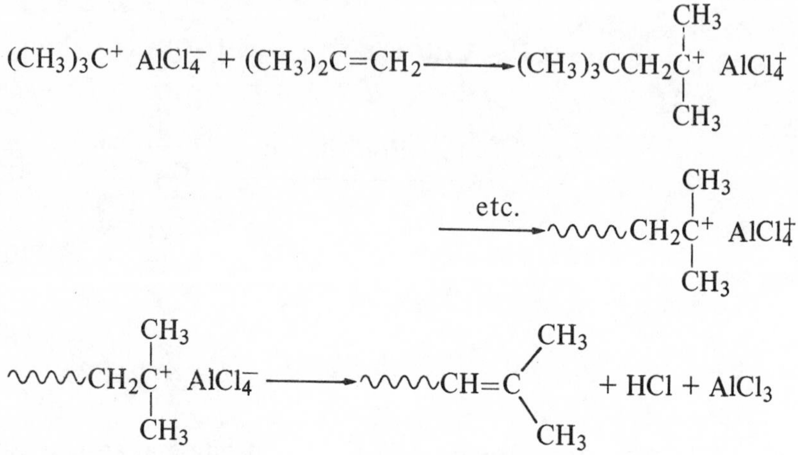

$$(CH_3)_3C^+ \ AlCl_4^- + (CH_3)_2C=CH_2 \longrightarrow (CH_3)_3CCH_2\overset{\overset{\displaystyle CH_3}{|}}{\underset{\underset{\displaystyle CH_3}{|}}{C}}{}^+ \ AlCl_4^-$$

$$\overset{etc.}{\longrightarrow} \mathrm{\sim\!\sim\!\sim} CH_2\overset{\overset{\displaystyle CH_3}{|}}{\underset{\underset{\displaystyle CH_3}{|}}{C}}{}^+ \ AlCl_4^-$$

$$\mathrm{\sim\!\sim\!\sim} CH_2\overset{\overset{\displaystyle CH_3}{|}}{\underset{\underset{\displaystyle CH_3}{|}}{C}}{}^+ \ AlCl_4^- \longrightarrow \mathrm{\sim\!\sim\!\sim} CH=C\begin{smallmatrix} \nearrow CH_3 \\ \\ \searrow CH_3 \end{smallmatrix} + HCl + AlCl_3$$

The isoprene comonomer provides unsaturated sites in the polymer so that it can be vulcanised.

The main use of butyl rubber is in tyre inner tubes. It is particularly well suited to this application since it has a very low permeability to gases.

Homopolymers of isobutene are also made and have a variety of applications.

During recent years, a major new use for isobutene has developed in the manufacture of **methyl *t*-butyl ether** (MTBE) for use as an octane number improver for gasoline. The use of MTBE, and other materials such as *t*-butyl alcohol, in gasoline is becoming of increasing importance as the use of lead additives is phased out.

MTBE is made by reacting isobutene and methanol in the presence of an acidic catalyst:

$$(CH_3)_2C=CH_2 + CH_3OH \longrightarrow (CH_3)_3COCH_3$$

The reaction is commonly carried out in the liquid phase in contact with an acidic ion exchange resin at temperatures in the range 40 to 90°C and pressures of about 10 atm.

5.5 *n*-BUTENES

The main chemical use of *n*-butenes is in the manufacture of *s*-**butyl alcohol** and **methyl ethyl ketone**.

In the manufacture of the alcohol, mixtures of *n*-butenes and butanes are used without separation. Hydration is brought about by sulphation followed by hydrolysis (cf. ethanol and isopropyl alcohol):

$$C_4H_8 \xrightarrow[\text{35--50°C}]{\text{75--85\% }H_2SO_4} s\text{-}Bu_2SO_4 + s\text{-}BuSO_4H$$

$$\xrightarrow[\text{100°C}]{H_2O} CH_3\overset{\displaystyle OH}{\underset{|}{C}}HCH_2CH_3$$

Some *s*-butyl alcohol is used in solvent applications. Most, however, is converted to methyl ethyl ketone, by dehydrogenation over a copper or a zinc oxide catalyst under conditions similar to those used for making acetone (section 4.2):

$$CH_3\overset{\displaystyle OH}{\underset{|}{C}}HCH_2CH_3 \longrightarrow CH_3\overset{\displaystyle O}{\underset{\|}{C}}CH_2CH_3 + H_2$$

The yield of methyl ethyl ketone based on butenes is about 80%.

Methyl ethyl ketone is extensively used as a solvent, in surface coatings and various other applications. US capacity is about 390 000 tonnes per annum.

Maleic anhydride can be made by oxidising *n*-butenes in the gas phase under conditions broadly similar to those used for oxidising benzene to maleic anhydride (section 7.5), and a certain amount is used for this purpose.

PROBLEMS AND EXERCISES

1. Approximately how much butadiene could be produced by a naphtha cracker making 500 000 tonnes per annum of ethylene?
2. Would you expect C_4 hydrocarbons to be important co-products

of ethylene manufacture in the Middle East?

3. In what respects is the butadiene-based chloroprene process described in section 5.3.2 inelegant?

4. Why is isobutene more reactive towards electrophiles than the n-butenes?

5. Suggest a mechanism for the reaction of isobutene with methanol.

6. Suggest routes that might be suitable for the large-scale manufacture of

   (i)  $CH_2CHCH_2CH_3$, from 1-butene;
            $\underset{O}{\diagdown \diagup}$

   (ii) 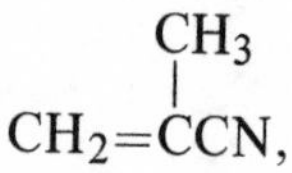 $\underset{\displaystyle CH_2=CCN}{\overset{\displaystyle CH_3}{\mid}}$, from isobutene.

6

Benzene, toluene and xylenes production

6.1 INTRODUCTION

Aromatic hydrocarbons have been produced commercially since the early part of the last century. Initially, they were simply used as mixtures, in solvent applications, but by 1850 benzene was being isolated as an individual compound.

The original source of aromatic hydrocarbons was **coal carbonisation.** Coal is carbonised for two main purposes; to provide coke for steel making and to produce coal gas, though this latter application is now of very little importance. A range of chemicals are given as by products, in rather small quantities. For example, benzene, toluene, and xylenes, the subjects of this chapter, are produced in the proportions indicated in Table 6.1.

Table 6.1
Benzene, toluene, and xylenes from coal carbonisation

Hydrocarbon	kg/tonne coal carbonised
Benzene	2–8
Toluene	0.5–2
Xylenes	0.1–0.5

Until around 1950 in the USA and 1960 in Western Europe, the quantities of benzene available from coal carbonisation were sufficient to meet the demand. Toluene had also been in adequate supply, apart from during the two world wars, and xylenes were not required as individual isomers until the 1950s. Over the past 25 to 35 years, however, the chemical industry has grown rapidly, while the amount of coal carbonised has decreased. Only a very small proportion of total demand for aromatic hydrocarbons can now be met from coal carbonisation products.

Most production of aromatic hydrocarbons is now from petroleum, via catalytic reforming and from the pyrolysis gasoline obtained as a co-product in naphtha and gas oil cracking. US capacity for benzene is about 7.7 million tonnes per annum.

6.2 CATALYTIC REFORMING

In the USA, catalytic reforming is the most important source of aromatic hydrocarbons. In Western Europe and Japan, it and production from pyrolysis gasoline are of approximately equal importance.

We have already briefly discussed the role of catalytic reforming as a refinery process and seen that, in refining, the naphtha fed to a reformer typically has a boiling range of about 75 to 200°C. In petrochemical operations, the naphtha used has a boiling range such that it contains mainly C_6 to C_8 compounds, that is, about 65 to 130°C. Apart from this, the process is very similar to that used in refining.

To avoid poisoning the reforming catalyst, it is necessary to remove sulphur-containing compounds from the naphtha. This is done by *hydrodesulphurisation*, that is, treating the naphtha with hydrogen over a cobalt oxide/molybdenum oxide catalyst at about 400°C. The sulphur is converted into hydrogen sulphide, which is readily separated.

6.2.1 Chemistry of reforming

In catalytic reforming, feed vapour, mixed with hydrogen in a ratio of about 6 moles hydrogen to 1 mole of hydrocarbon, is reacted over a catalyst at temperatures and pressures in the range 450 to 550°C and 10 to 35 atm. The catalyst is a *dual function catalyst*, consisting of platinum, often promoted with rhenium, on a support of halogen-treated alumina. It provides a combination of hydrogenation/dehydrogenation activity (the platinum) and acidity (the alumina).

The aromatic hydrocarbons are formed via three main reactions:

(i) *dehydrogenation of cyclohexane and homologues*, e.g.

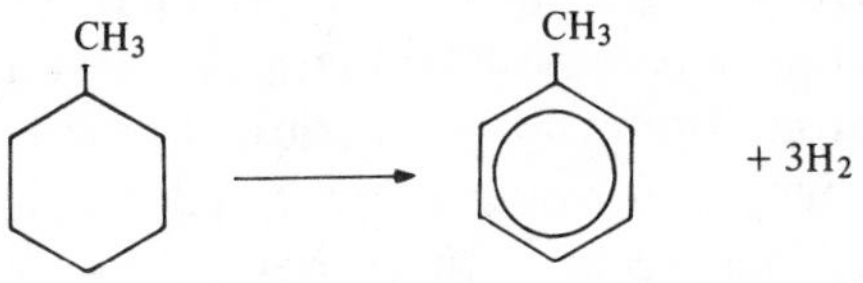

$+ 3H_2$

(ii) *dehydroisomerisation of cyclopentane homologues*, e.g.

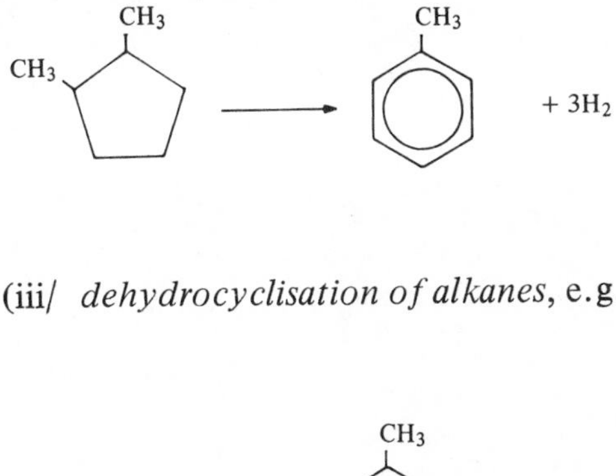

$+ 3H_2$

(iii/ *dehydrocyclisation of alkanes*, e.g.

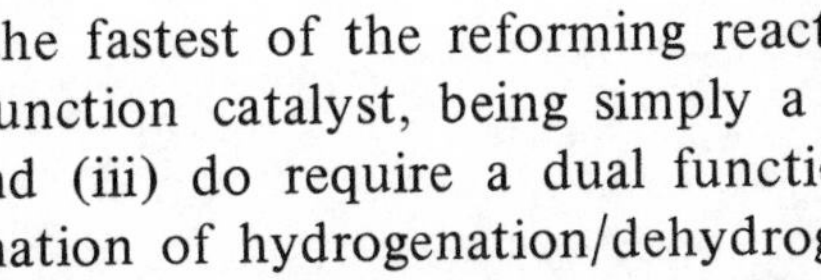

C_7H_{16} $+ 4H_2$

Reaction (i) is the fastest of the reforming reactions. It does not require a dual function catalyst, being simply a dehydrogenation. Reactions (ii) and (iii) do require a dual function catalyst. They involve a combination of hydrogenation/dehydrogenation reactions and carbocation rearrangements, catalysed by platinum sites and acidic sites respectively. Their mechanisms are complicated, and are by no means fully understood. However, provided that it is borne in mind that the schemes are much simplified, and that the reactive intermediates are adsorbed on the catalyst, they can be visualised as following pathways of the type indicated in Fig. 6.1.

The dehydrocyclisation reactions are much slower than the dehydrogenation and dehydroisomerisation reactions. Also, hydrocracking of alkanes occurs at an appreciable rate, e.g.

$$C_7H_{16} + H_2 \longrightarrow C_3H_8 + C_4H_{10}$$

Consequently, naphthas containing high proportions of cycloalkanes are preferred as feeds for aromatics production.

The aromatic-forming reactions are reversible. One might, therefore, ask why reforming is operated at elevated pressure and with added hydrogen, since these conditions will have an unfavourable

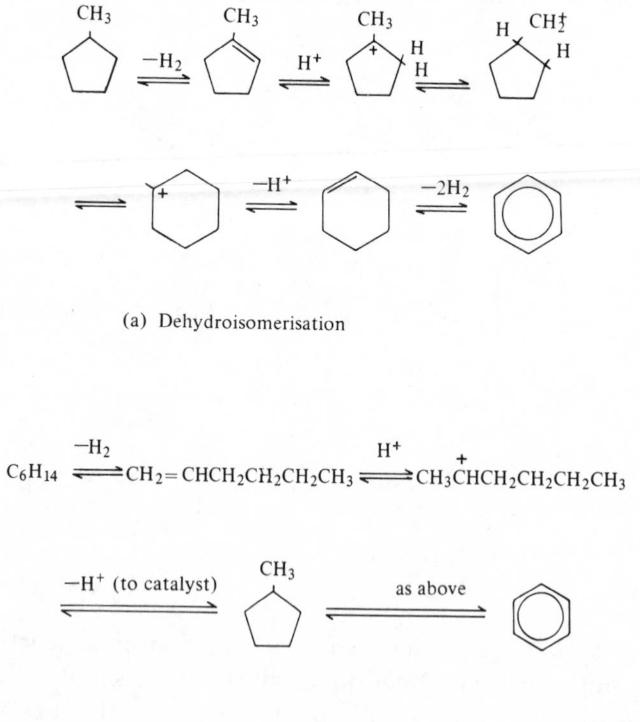

(a) Dehydroisomerisation

(b) Dehydrocyclisation

Fig. 6.1 – Catalytic reforming mechanisms (simplified and idealised).

effect on the position of the equilibrium. The answer is that this is done in order to limit coke deposition on the catalyst. In many gas phase catalytic reactions, high molecular weight carbonaceous material, called coke, tends to be formed via a combination of polymerisation, cyclisation, and dehydrogenation reactions. It deposits on the catalyst, and destroys or reduces its activity. The problem is dealt with in various ways in different processes. In catalytic reforming, the rate of coke deposition is reduced by operating under a high partial pressure of hydrogen. This does not, however, totally eliminate coke formation, and the catalyst has to be periodically regenerated.

6.2.2 Reaction systems

Catalytic reforming is, overall, an endothermic reaction, and so heat has to be supplied. The design of the reaction system is largely determined by this requirement and by the necessity to provide for catalyst regeneration. A variety of systems are used. All essentially comprise a number of adiabatic reactors, typically three or four, in series, with provision for heat input to the process stream between reactors.

Catalyst regeneration is brought about by burning off the coke under carefully controlled conditions with a gas stream containing from 0.5% to 1% of oxygen. Most frequently this is done with the catalyst *in situ* in the reactors. In some processes the whole plant is shut down periodically, e.g. every 12 months, for catalyst regeneration. In others an extra 'swing' reactor is provided, so that at any time the catalyst in one reactor can be undergoing regeneration without stopping production. In some of the more modern plants, regeneration is carried out continuously. For example, in one process the reactors are vertically stacked. Regenerated catalyst is fed to the top reactor, and catalyst moves down through the stack by gravity as a small quantity is continuously removed from the bottom reactor for regeneration and recycle.

6.2.3 Product separation

We have already noted that the major reforming reactions are reversible. Under the process conditions used, conversion to aromatic hydrocarbons is by no means complete, so that the reformate contains unconverted alkanes and, to a lesser extent, cycloalkanes. This presents difficulties in product isolation, since some of these components have boiling points very close to, or identical with, those of the aromatic hydrocarbons. Separation simply by fractional distillation is therefore impossible.

The molecules of aromatic hydrocarbons, by virtue of their π-electrons, are much more polarisable than are those of alkanes and cycloalkanes. In consequence, aromatic hydrocarbons are soluble in a variety of polar solvents which do not dissolve alkanes and cyclo-alkanes, and they can be separated from alkanes and cycloalkanes by solvent extraction. This is done by passing the reformate counter-current to solvent in an apparatus which brings about good contact between the two immiscible phases. One type of equipment used is a *rotating disc contactor*, a vertical tower through which passes a rotating shaft bearing a number of discs. Solvent is passed into the top of the tower, and reformate into the bottom. The solvent phase, which has a higher specific gravity than the reformate, moves down the tower, and the reformate moves up it. By the time the solvent reaches the bottom of the tower, it has extracted most of the aro-matics content of the reformate. The residual alkanes and cyclo-alkanes pass out of the top of the contactor. The aromatic hydro-carbons are stripped from the extract and benzene, toluene, and a mixed C_8 stream are separated by fractional distillation. Fig. 6.2 shows an aromatics separation plant in schematic form.

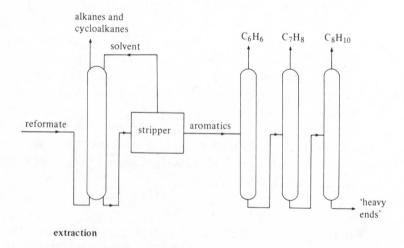

Fig. 6.2 – Aromatics separation plant.

A variety of solvents can be used, and a number of proprietary separation processes exist. The most commonly used solvents at present are tetraethylene glycol and sulfolane:

$$HOCH_2CH_2(OCH_2CH_2)_2OCH_2CH_2OH$$

tetraethylene glycol

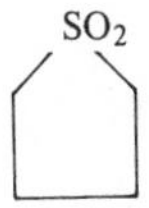

sulfolane

The proportions of benzene, toluene, and C_8 hydrocarbons typically produced in catalytic reforming are shown in Table 6.2.

6.3 AROMATICS FROM PYROLYSIS GASOLINE

As we saw in Chapter 2, when ethylene is made by cracking naphtha or gas oil, a fraction with a boiling range of about 20 to 200°C, called pyrolysis gasoline, is obtained. This contains a high proportion of aromatic hydrocarbons, together with alkenes, dienes, alkanes and cycloalkanes. It may be used as motor gasoline, but is also a valuable source of aromatic hydrocarbons.

Isolation of benzene, toluene, and C_8 aromatics from pyrolysis gasoline is carried out in the same way as from reformate, but with one important difference. The alkenes and dienes are soluble in the extraction solvents used, and have to be removed before the extraction stage. This is done by treating the pyrolysis gasoline with hydrogen under conditions such that the alkenes and dienes are hydrogenated but the aromatic hydrocarbons are unaffected. For example, one process uses nickel or palladium catalysts, and temperatures and pressures from 75 to 150°C and 10 to 40 atm. After hydrogenation, the gasoline is distilled to give a fraction containing mainly C_6 to C_8 hydrocarbons, and benzene, toluene and C_8 aromatics are separated as described above. A typical product distribution from pyrolysis gasoline is shown in Table 6.2.

Table 6.2

Aromatic hydrocarbons:
Typical product distributions and market demand (%)

	Reformate	Pyrolysis gasoline	Market demand (USA)
Benzene	11	54	60
Toluene	55	31	6
C_8s	34	15	34

6.4 C_8 AROMATICS

The C_8 streams obtained as described above contain o-, m-, and p-xylenes in the approximate proportions indicated in Table 6.3. o-Xylene and p-xylene are required in large quantities for the manufacture of phthalic anhydride and terephthalic acid respectively; the demand for m-xylene is much smaller, and most plants do not separate it. Ethylbenzene is also a major petrochemical intermediate, but is mostly made by synthesis from ethylene and benzene (section 7.1.1).

Table 6.3
Aromatic C_8 stream

	Typical composition (wt %)		b.p. (°C)	m.p. (°C)
	Catalytic reformate	Pyrolysis gasoline		
o-Xylene	24	12	144	−25
m-Xylene	40	25	139	−48
p-Xylene	19	10	138	13
Ethylbenzene	17	53	136	−95

It is clear from Table 6.3 that the C_8 stream from catalytic reforming is a better feed for the production of xylenes than that from pyrolysis gasoline.

6.4.1 Separation

As would be expected, the boiling points of the C_8 aromatic hydro-carbons are very close together (see Table 6.3). *o*-Xylene can be separated from the other components by fractional distillation, without any major problems. Separation of ethylbenzene in this way is also technically feasible, but, as a consequence of the small difference in boiling points between ethylbenzene and *m*- and *p*-xylenes, is very expensive. Separation of *m*-xylene and *p*-xylene by distillation is not possible.

The original method used to separate *p*-xylene was fractional crystallisation at low temperatures. It can be seen from Table 6.3 that the melting point of *p*-xylene is much higher than that of any of the other C_8 hydrocarbons. If a mixture of C_8s is cooled to about −60°C, *p*-xylene crystallises out and can be filtered off. Recrystallisation, again at low temperatures, gives a product of acceptable purity. About 60% to 70% of the *p*-xylene content of the mixture can be recovered in this way. Operating at these low temperatures is expensive, both in terms of capital cost and operating cost.

Since the early 1970s, an alternative method of separation has been coming into increasing use. This involves adsorption of *p*-xylene onto a zeolite adsorbent which does not take up the other C_8 hydrocarbons, followed by desorption.

Combination of fractional distillation to give *o*-xylene, followed by separation of *p*-xylene by either adsorption or crystallisation, gives a xylenes plant with the configuration indicated in Fig. 6.3.

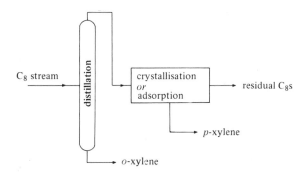

Fig. 6.3 — Basic xylenes plant.

6.4.2 Isomerisation

A xylenes plant of the type just described produces large amounts of a residual C_8 stream, consisting mainly of *m*-xylene and ethylbenzene. This may be disposed of for use as a solvent or for gasoline blending. However, it is common to put it through an isomerisation stage to produce further *o*- and *p*-xylenes.

The isomerisation processes which were first used involve contacting the C_8s with a strongly acidic catalyst. Typically the reaction is carried out in the vapour phase at about 500°C over a silica–alumina catalyst. Xylenes are protonated to *areneonium ions*, which isomerise by 1,2-hydride and methide shifts, e.g.:

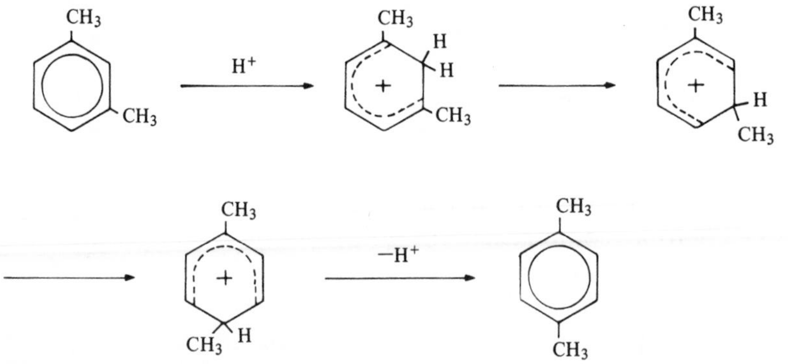

These reactions occur extremely rapidly under the process conditions, to give essentially an equilibrium mixture of xylenes, with the approximate composition 20% *o*-xylene, 55% *m*-xylene, and 25% *p*-xylene.

Processes of the type just described do not convert ethylbenzene into xylenes. However, there are a number of processes which will do this. These mainly employ dual function catalysts similar to those used in reforming, and the reaction is carried out in the presence of hydrogen. Isomerisation is believed to occur via mechanisms similar to those involved in catalytic reforming.

A variety of combinations of isomerisation and isomer separation are used. Maximum output of *o*- and *p*-xylenes is achieved by use of an isomerisation stage which will convert ethylbenzene into xylenes, in a scheme as indicated in Fig. 6.4. With such a plant, variation in relative output of the two isomers to suit market requirements is simply achieved. If, for example, it is desired to reduce the output of *o*-xylene, all that is necessary is to reduce the offtake from the fractional distillation.

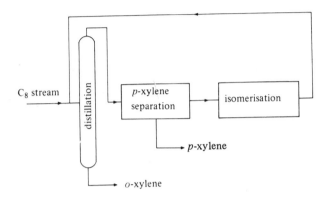

Fig. 6.4 – Xylenes plant with isomerisation stage.

6.5 HYDRODEALKYLATION AND DISPROPORTIONATION

We have seen that the product distribution from catalytic reforming does not correspond at all well with the market demand for aromatic hydrocarbons. Consequently, where a major proportion of benzene is produced by catalytic reforming, for example, in the USA, the amount of toluene available is far in excess of requirements for chemical applications. This excess can readily be disposed of as a gasoline blending component, but the return obtained from this varies widely with the state of the gasoline market. Alternative outlets are *hydrodealkylation*, which converts the toluene to benzene, and *disproportionation*, which converts it to benzene and xylenes.

Hydrodealkylation involves treatment of toluene with hydrogen:

Both catalytic and non-catalytic processes are used. In the former, the catalysts are mixed metal oxides, e.g. of aluminium, molybdenum and cobalt, and reaction is carried out at 550 to 650°C and 35 to 70 atm. Without a catalyst, temperatures used are about 100°C higher, but otherwise the conditions are similar. In both cases, the yields are around 99%.

The economic attractiveness of hydrodealkylation is determined by the difference between the price of benzene and of toluene, which in turn is determined by the state of the gasoline market. It is normal to operate hydrodealkylation plants when the gap between the prices is high and close them down when it is low.

Disproportionation is carried out over acidic catalysts, e.g. over a zeolite at around 480°C. It gives benzene and a mixture of *o*-, *m*-, and *p*-xylenes which can be separated as discussed in section 6.4.1. The process is operated to a much smaller extent than hydrodealkylation.

PROBLEMS AND EXERCISES

1. Why did toluene go into short supply during the two world wars? (Hint: Look at the table of contents for Chapter 8.)
2. During catalytic reforming, any sulphur and nitrogen compounds present are converted into hydrogen sulphide and ammonia respectively. The hydrogen sulphide poisons the catalyst by occupying platinum sites. How would you expect the ammonia to affect the catalyst? To what extent would you expect aromatic hydrocarbons to be formed over a catalyst poisoned by (i) hydrogen sulphide and (ii) ammonia?
3. What is the origin of the aromatic hydrocarbons in pyrolysis gasoline?
4. Write down the steps involved in the isomerisation of *m*-xylene to *o*-xylene over an acidic catalyst.
5. Draw an outline flow scheme for a plant to produce *p*-xylene only from a C_8 reformate stream.
6. How important do you think pyrolysis gasoline is likely to be as a source of aromatic hydrocarbons in the Middle Eastern petrochemical industry?

7

Benzene derivatives

Benzene is an intermediate with a similar breadth of applications to ethylene, and is by far the most important of the aromatic hydrocarbons. We shall discuss its major derivatives roughly in order of commercial importance.

7.1 ETHYLBENZENE AND STYRENE

Styrene was first produced commercially in the USA and Germany in the 1930s. The scale of operation was initially modest, but during World War II, particularly in the USA, production of styrene was increased at an enormous rate, for use in the manufacture of synthetic rubber.

We saw in section 4.3 that since 1968 styrene has been made as a co-product of propylene oxide by the Halcon process. This route is increasing in importance. However, the main route to styrene is by catalytic dehydrogenation of ethylbenzene, prepared by alkylation of benzene with ethylene:

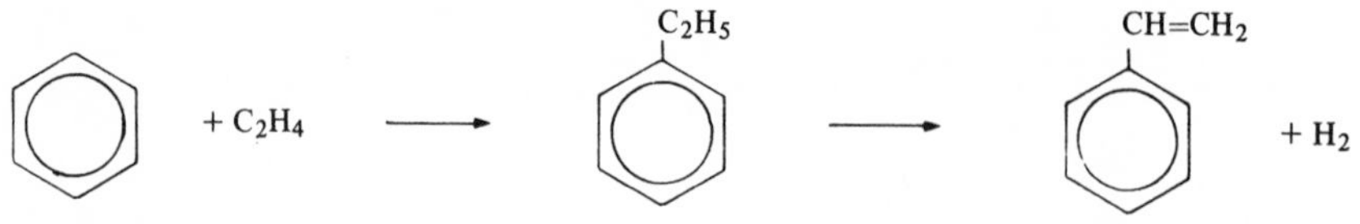

Manufacture of styrene consumes about 50% of benzene.

7.1.1 Ethylation

Until fairly recently this was almost exclusively carried out in the liquid phase, in the presence of an aluminium chloride catalyst, with either hydrogen chloride or ethyl chloride as promoter. Temperatures used range from about 80 to 130°C, and pressures from atmospheric pressure to about 9 atm. The reactivation is a typical Friedel-Crafts reaction. Its mechanism may be envisaged as follows:

$$CH_2{=}CH_2 + AlCl_3 + HCl \longrightarrow C_2H_5^+ \ AlCl_4^-$$

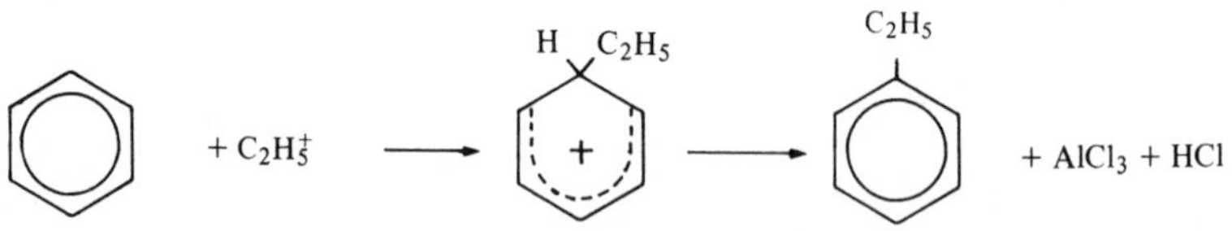

Ethylbenzene is substantially more readily attacked by electro-philes than benzene, and consequently polyalkylation tends to occur. To reduce the extent of this, benzene is used in excess. Typically a molar ratio of benzene to ethylene of about 1.7:1 is used. Some polyethylbenzenes are still formed, but these are recycled to the reactor, where they undergo transalkylation reactions with benzene:

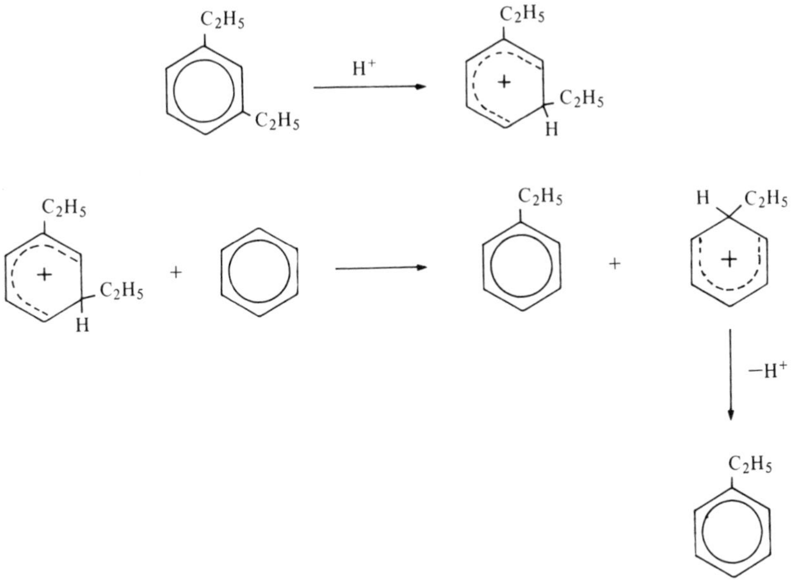

Overall, there is no production of polyethylbenzenes, and the yield of ethylbenzene is very high, at around 98%.

The main problem with this process is that the reaction mix is very corrosive, and this gives rise to some difficulties in plant construction and maintenance. Operation in the gas phase over a solid catalyst avoids these difficulties, and processes using silica–alumina and solid phosphoric acid catalysts have been available for many years. However, these suffer from the drawback that the catalysts are not strong enough acids to catalyse transalkylation, and they have been used to only a limited extent.

In the last few years the situation has been transformed by the introduction of a new gas phase process, the Mobil–Badger process, in which the catalyst is a zeolite which also catalyses transalkylation. Reaction is carried out at about 420°C and 12 to 20 atm., with an excess of benzene. As in the liquid phase process, benzene, ethylbenzene, and polyethylbenzenes are separated by distillation, and the benzene and polyethylbenzenes are recycled to the reactor. The yield is again about 98%. This appears to be the process of choice for new plants, though the liquid phase process continues to be operated in many existing plants.

7.1.2 Dehydrogenation

Ethylbenzene is dehydrogenated in the gas phase over catalysts based on iron oxide. The reaction is reversible and endothermic. High temperatures are thus favourable both to the rate of reaction and to the position of the equilibrium. However, a limit is imposed on the temperature used by the fact that above about 610°C thermal cracking of ethylbenzene begins to occur at a significant rate, and results in loss of yield.

There are two major versions of the dehydrogenation process. In that first developed in the USA, the reaction is carried out in adiabatic reactors, whereas the version first developed in Germany uses tubular reactors.

In the **adiabatic process**, the ethylbenzene is preheated to about 500°C and is then mixed with superheated stream at about 800°C immediately before entering the catalyst bed. The resulting mixture of steam and ethylbenzene (molar ratio steam to ethylbenzene about 14:1) enters the bed at about 650°C. As the gas stream passes through the bed the temperature drops, owing to the endothermicity of the reaction, and the gases emerge at about 590°C. The conversion is around 40% and the yield after recycle about 90%. The use of two or more reactors in series with interstage heating gives better yields and/or conversions, but at the expense of higher capital costs.

In the **tubular reactor process**, the catalyst is held in tubes heated in a furnace. The steam/ethylbenzene mixture enters the tubes at around 580°C and the exit temperature is about 610°C. The conditions are thus more favourable than those in the adiabatic process both in terms of the equilibrium position at the exit temperature, and the exposure to temperatures which bring about thermal cracking, and higher conversions and yields are achievable, at lower steam ratios. However, the capital cost of a tubular reactor is much higher than that of an adiabatic reactor, and the adiabatic process is much more widely used.

Separation of styrene from the reaction mixture presents a certain amount of difficulty, since styrene readily polymerises thermally. Vacuum distillation is used to minimise the temperatures, and a polymerisation inhibitor, e.g. sulphur, is added to the mixture before distillation. Small amounts of benzene and toluene, formed in the cracking reactions, are obtained as by-products.

US capacity for styrene is about 4 million tonnes per annum.

7.1.3 Uses of styrene

Virtually all styrene is used in the manufacture of polymers. Polystyrene, which consumes about 63% of styrene, is made in three types, all produced by free radical polymerisation. Polymerisation of styrene alone gives *crystal polystyrene*, a transparent, glass-like polymer. Polymerisation of styrene containing synthetic rubber in solution gives *toughened polystyrene*, an opaque material with much higher resistance to impact than crystal polystyrene. *Expandable polystyrene* is made by suspension polymerisation of styrene containing small amounts of an inert volatile hydrocarbon such as pentane. It is initially in the form of small beads. If these are heated with steam, the polystyrene softens, and the pentane volatilises, blowing the beads up like popcorn. The familiar polystyrene foam, extensively used in packaging and insulation, is made by carrying out this procedure in a mould, so that the expanded beads fuse together.

SBR rubber (section 5.3.1) consumes about 16% of styrene, unsaturated polyesters (section 8.4.1) about 6%, and the rest is used in a variety of copolymers.

7.2 CYCLOHEXANE AND NYLON INTERMEDIATES

The importance of cyclohexane as a chemical intermediate derives almost entirely from its use in the production of adipic acid and hexamethylene diamine for nylon 66, and caprolactam for nylon 6. Manufacture of cyclohexane currently consumes about 15% of benzene.

7.2.1 Cyclohexane

Cyclohexane is made by the catalytic hydrogenation of benzene:

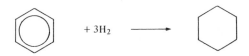

This reaction is reversible; we have already seen that dehydrogenation of cyclohexane and its homologues is one of the important aromatic-foming reactions in catalytic reforming. In the direction of hydrogenation the reaction is exothermic, so that the equilibrium constant decreases with increase in temperature. Thus, high equilibrium convesion of benzene to cyclohexane is favoured by low temperatures and high partial pressures of hydrogen. At around 250°C and under fairly moderate pressures, conversion of benzene at equilibrium is almost 100%. With an appropriate choice of catalyst and conditions the yield is also virtually 100%, so that this process is unusual in that product may be obtained direct from the reaction system with no need for purification.

A number of processes are operated, differing in the type of catalyst, in whether the reaction is carried out in the liquid or the gas phase, and in the arrangement of the reactor system. Catalysts are usually based on nickel or platinum. Typically in a gas phase process there are a number of adiabatic reactors in series with inter-stage cooling to remove the heat of reaction. An excess of hydrogen is used, at a pressure of about 30 atm., and reaction temperatures are up to about 350°C in some of the reactors. The temperature in the last reactor determines the concentration of benzene in the product. It is typically in the range 170 to 250°C.

US capacity for cyclohexane is about 1.4 million tonnes per annum.

7.2.2 Cyclohexanol and cyclohexanone

Most cyclohexane is oxidised to a mixture of cyclohexanol and cyclo-hexanone, as a first step in the manufacture of nylon intermediates. A small proportion is used for the manufacture of caprolactam by a different route (section 7.2.6).

The oxidation of cyclohexane is a further example of the use of liquid phase free radical oxidation in chemical manufacture (cf. sections 3.6.1 and 4.3). Typically, in what may be called the con-ventional process, oxidation is carried out at 150 to 160°C and around 10 atm., using air as the oxidising gas, and with a soluble cobalt salt as catalyst. Three or four stirred-flow reactors in series are usually used. Conversion is limited to about 4%, for reasons that will become clear shortly.

The initial reaction product in the oxidation is cyclohexyl hydro-peroxide, produced by steps of the type discussed in section 4.3:

$$C_6H_{12} \xrightarrow{\ -[H\cdot]\ } C_6H_{11}\cdot \xrightarrow{\ O_2\ } C_6H_{11}OO\cdot \xrightarrow{\ RH\ } C_6H_{11}OOH + R\cdot$$

Under the reaction conditions used, the hydroperoxide rapidly under-goes further reaction, partially thermally as we saw in section 4.3, but mainly by much faster one electron transfer reactions involving the catalyst:

$$C_6H_{11}OOH + Co^{2+} \longrightarrow C_6H_{11}O\cdot + OH^- + Co^{3+}$$

$$C_6H_{11}OOH + Co^{3+} \longrightarrow C_6H_{11}OO\cdot + H^+ + Co^{2+}$$

We can now write down steps which lead to the formation of cyclohexanol and cyclohexanone:

$+ RH \longrightarrow + R\cdot$

$+ X\cdot \longrightarrow + XH$

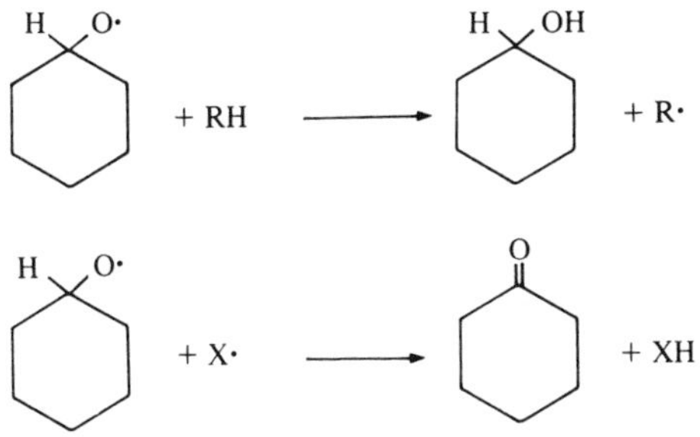

There are, however, many other reactions which can occur in the system. For example, the cyclohexyloxy radical can undergo a ring opening β-scission reaction, and both cyclohexanol and cyclohexanone can undergo further oxidation:

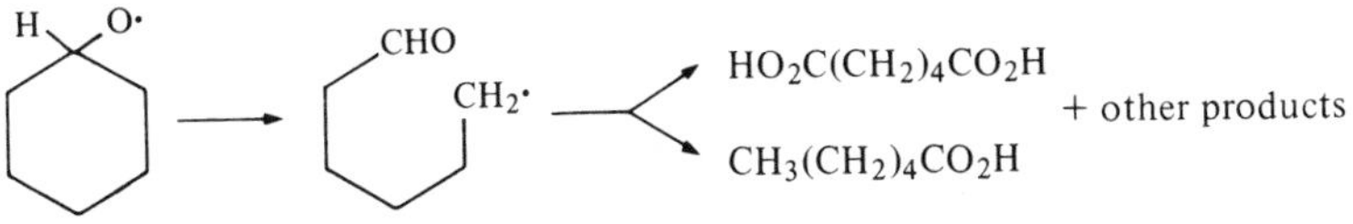

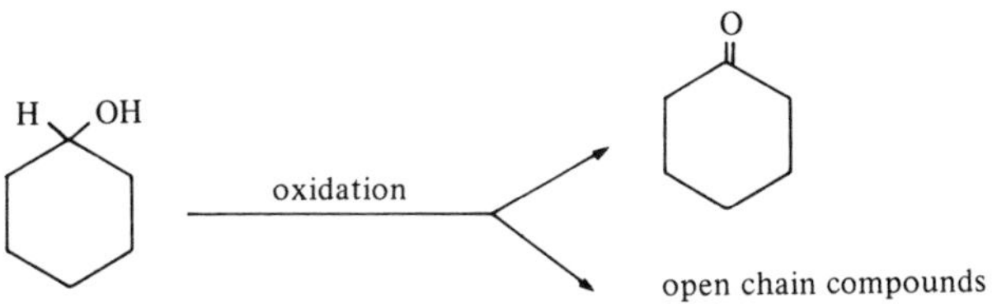

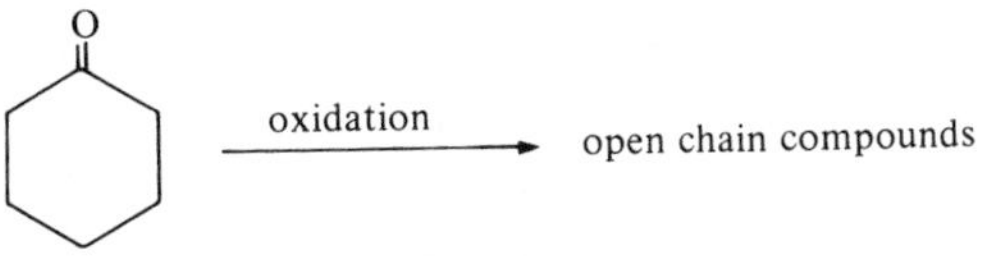

$\longrightarrow$ CHO, $CH_2\cdot \longrightarrow HO_2C(CH_2)_4CO_2H$ + other products $CH_3(CH_2)_4CO_2H$

oxidation $\longrightarrow$ open chain compounds

oxidation $\longrightarrow$ open chain compounds

It is in order to reduce the extent of this further oxidation that the process is operated at low conversions.

The product from the oxidation reactors contains, in addition to cyclohexane, cyclohexanol and cyclohexanone, mono- and dibasic acids and their cyclohexyl esters. It is treated with aqueous sodium hydroxide to hydrolyse the esters, and then cyclohexane and a mixture of cyclohexanol and cyclohexanone are separated by distillation. The cyclohexane is of course recycled. The yield of cyclohexanol/cyclohexanone, in a molar ratio of approximately 2:1, is 75% to 80%.

During the 1960s an interesting variant of the process was introduced, in which the oxidation is carried out in the presence of boric acid and with no cobalt catalyst. The main oxidation product in this case is cyclohexyl borate, together with a certain amount of cyclohexanol and cyclohexanone. The product is treated with water to hydrolyse the ester to cyclohexanol, and boric acid, which is recycled. At a given conversion, this process gives a significantly higher yield than the conventional process. Thus, operated at a conversion of 5% it gives a yield of 85% to 90%. The ratio of cyclohexanol to cyclohexanone is about 9:1, much higher than in the conventional process.

The changes in yield and product distribution are thought to be due to two factors. Firstly, most of the cyclohexyl hydroperoxide reacts directly with boric acid to give the cyclohexanol ester:

$$C_6H_{11}OOH \xrightarrow{H_3BO_3} (C_6H_{11}O)_3B$$

Since cyclohexyloxy radicals are not involved in this reaction, loss of yield by ring opening reactions does not occur. Secondly, cyclohexyl borate is much more stable to oxidation than cyclohexanol and cyclohexanone.

Although this process has a significant advantage over the conventional process in terms of yield and/or conversion, the increased process complexity which results from the extra reaction stage and the handling and recycling of boric acid to some extent at least cancels this out. Both processes are used.

More recently, a further variant has been developed in the so-called high peroxide process. In this, the conditions used for the oxidation are such that the hydroperoxide is the main product. This is then catalytically decomposed under controlled conditions. It is not clear to what extent this process is operated commercially.

Cyclohexanol and cyclohexanone are also made by the hydrogenation of phenol (section 7.3.3) but this route is now of only relatively minor importance.

7.2.3 Adipic acid

Adipic acid is made by oxidising cyclohexanol/cyclohexanone mixture with nitric acid:

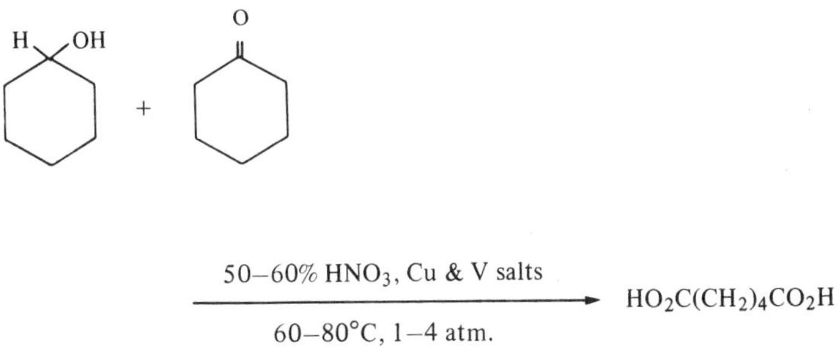

$$\text{50–60\% HNO}_3, \text{Cu \& V salts} \atop \text{60–80°C, 1–4 atm.} \longrightarrow HO_2C(CH_2)_4CO_2H$$

Yields of 92% to 96% are obtained, with complete conversion of the cyclohexanol and cyclohexanone. The adipic acid is purified by recrystallisation from water.

It is evident that it would be preferable to use air for this oxidation, or better still, to make adipic acid in one stage by air oxidation of cyclohexane. From our discussion of the mechanism of the oxidation of cyclohexane, it can be seen that it is intrinsically difficult to obtain substantial yields of adipic acid, in consequence of the multiplicity of reaction paths open to the free radical intermediates. Although a great deal of effort has been expended in trying to develop air oxidation processes, at present the nitric acid process appears to be almost universally used.

More than 90% of adipic acid is consumed in the manufacture of nylon 66. In this, it is used both as such and as an intermediate in the manufacture of hexamethylene diamine.

7.2.4 Adiponitrile and hexamethylenediamine

Conversion of adipic acid to adiponitrile is brought about by reacting it with ammonia in the presence of a dehydrating catalyst:

$$HO_2C(CH_2)_4CO_2H + 2NH_3 \longrightarrow NC(CH_2)_4CN + 4H_2O$$

Both liquid and gas phase processes are used. For example, the reaction may be carried out in the gas phase at about 350°C over a boron phosphate catalyst. Yields of about 90% are obtained.

As we have already seen, adiponitrile is also made from acrylonitrile (section 4.8) and from butadiene (section 5.3.3).

Hexamethylenediamine is produced by hydrogenation of adiponitrile:

$$NC(CH_2)_4CN + 4H_2 \longrightarrow H_2N(CH_2)_6NH_2$$

The hydrogenation is carried out in the presence of ammonia, which has the effect of reducing the extent of side reactions. A variety of catalysts and conditions are used. For example, the reaction may be carried out over an iron catalyst at 150°C and 300 atm. Yields of 95% or more are obtained.

7.2.5 Nylon 66

Nylon 66, the first of the synthetic fibres, was invented by Carothers, of du Pont, in the mid 1930s. It was first put into commercial production in 1938.

Production of nylon 66 involves two steps from adipic acid and hexamethylenediamine. In the first, *nylon salt* is made by mixing equimolar quantities of hexamethylenediamine and adipic acid:

$$H_2N(CH_2)_6NH_2 + HO_2C(CH_2)_4CO_2H$$
$$\longrightarrow H_3^+N(CH_2)_6NH_3^+ \ ^-O_2C(CH_2)_4CO_2^-$$

The salt may be isolated as a solid by mixing the two reagents in solution in methanol, when it precipitates out, or may simply be prepared in aqueous solution. The former method was useful in the early days of nylon production, since the solid salt could be purified, thus ensuring an exact balance of functional groups for the polymerisation. Now that high-purity adipic acid and hexamethylene-diamine are available, it is probably not necessary.

Polymerisation is brought about by heating a concentrated aqueous solution of nylon salt at temperatures rising to 275°C:

$$H_3^+N(CH_2)_6NH_3^+ \ ^-O_2C(CH_2)_4CO_2^-$$
$$\longrightarrow \ +HN(CH_2)_6NHOC(CH_2)_4CO+_n \ + 2nH_2O$$

Initially the reaction is carried out under pressure, but as the polymerisation proceeds the pressure is reduced to allow the removal of water, as steam.

To control the molecular weight, a small amount of acetic acid is added at the start of the polymerisation. This acts by 'end capping' some of the polymer molecules.

The main importance of nylon 66 is as a synthetic fibre, but it is also used to some extent as a plastic.

7.2.6 Caprolactam

Most caprolactam is made from cyclohexanone, though other routes are also in commercial use. Originally the cyclohexanone was made by hydrogenating phenol (section 7.3.3), but material prepared from cyclohexanol/cyclohexanone mixtures produced as described above is now also commonly used. In this case, the mixture is passed over a zinc oxide catalyst at temperatures in the range 375 to 425°C, when most of the cyclohexanol is converted into cyclohexanone. The cyclohexanone is separated by distillation, and the remaining cyclohexanol is recycled.

Cyclohexanone is converted into caprolactam via a Beckmann rearrangement of its oxime:

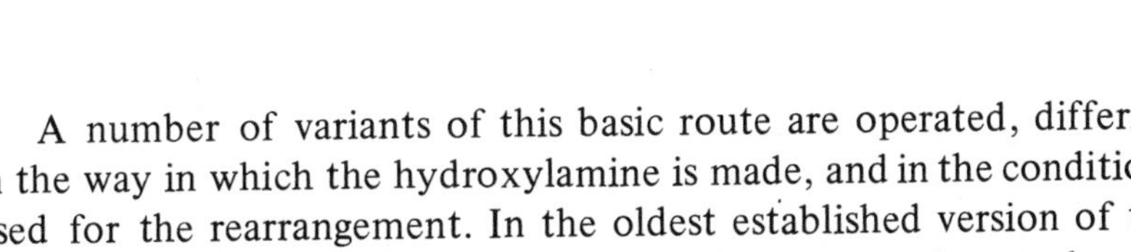

A number of variants of this basic route are operated, differing in the way in which the hydroxylamine is made, and in the conditions used for the rearrangement. In the oldest established version of the process, hydroxylamine is made from ammonium carbonate by the Raschig process:

$$(NH_4)_2CO_3 \xrightarrow{\text{NO, NO}_2} NH_4NO_2 \xrightarrow{\text{NH}_3,\text{SO}_2,\text{H}_2\text{O}} HON(SO_3NH_4)_2$$

$$\xrightarrow{\text{H}_2\text{O}} NH_2OH \cdot H_2SO_4 + (NH_4)_2SO_4$$

The hydroxylamine sulphate, together with excess sulphuric acid, is reacted with cyclohexanone at around 100°C, in the presence of ammonia:

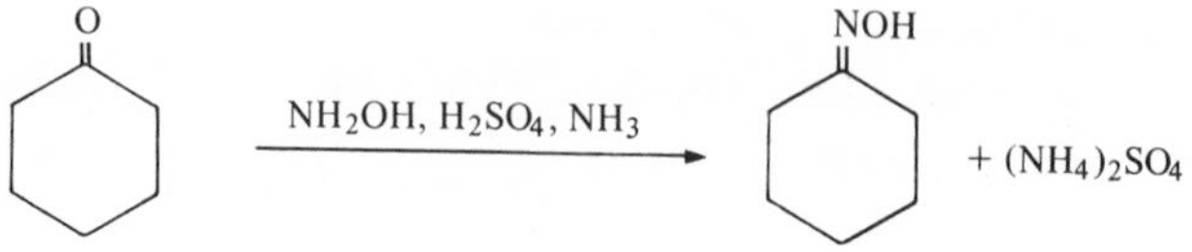

The oxime is then treated with oleum (sulphur trioxide dissolved in sulphuric acid) at about 120°C to bring about the Beckmann rearrangement, and the reaction mix is neutralised with ammonia. The caprolactam is purified by crystallisation or vacuum distillation. The yield is about 95%.

Ammonium sulphate is produced at each stage of this process, and in all about 4.5 tonnes per tonne of caprolactam is obtained. The return that can be obtained for this, by selling it for use as a fertiliser, has a profound effect on the economics of the process. There have been periods when this return has been very small, because of lack of demand, and consequently there has been a lot of research activity directed towards the development of versions of the process which do not produce so much ammonium sulphate. A number of such versions are operated, as are processes following other routes, for example, one based on toluene (section 8.2).

US capacity for caprolactam (all processes) is about 530 000 tonnes per annum.

7.2.7 Nylon 6

Nylon 6 was first produced in Germany in 1940. It is made by polymerising caprolactam in the presence of a catalytic amount of water, under conditions similar to those used in nylon 66 manufacture. The water hydrolyses a small amount of caprolactam to ϵ-aminocaproic acid, and then a combination of ring opening reactions involving $-NH_2$ groups, and condensation reactions between $-NH_2$ groups and $-CO_2H$ groups leads to the build up of polymer molecules:

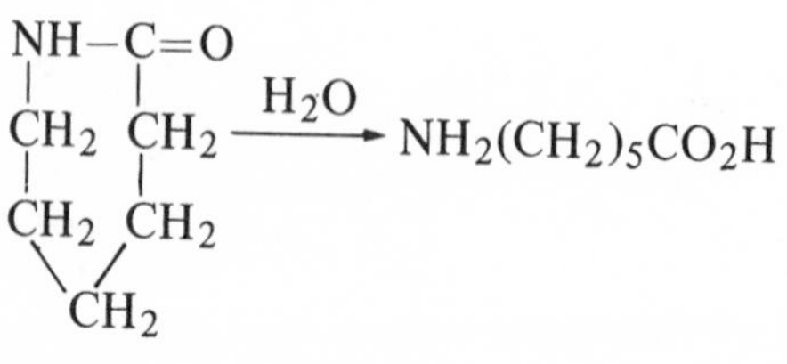

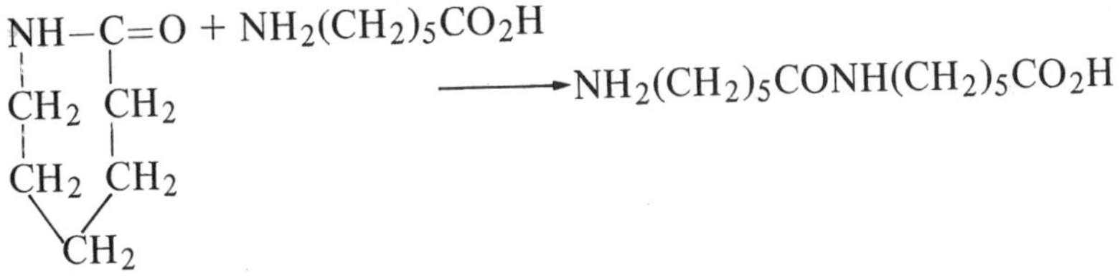

$$NH-C=O + NH_2(CH_2)_5CO_2H$$
$$\underset{|}{CH_2}\ \underset{|}{CH_2} \qquad\qquad \longrightarrow NH_2(CH_2)_5CONH(CH_2)_5CO_2H$$
$$\underset{|}{CH_2}\ \underset{|}{CH_2}$$
$$CH_2$$

$$\xrightarrow{\text{etc.}} NH_2(CH_2)_5CO[NH(CH_2)_5CO]_nOH$$

$$\sim\!\sim\!\sim NH(CH_2)_5CO_2H + NH_2(CH_2)_5CO\sim\!\sim\!\sim$$

$$\longrightarrow \sim\!\sim\!\sim NH(CH_2)_5CONH(CH_2)_5CO\sim\!\sim\!\sim + H_2O$$

Overall, the polymerisation is reversible, and the equilibrium constant is such that at the end of the polymerisation the polymer contains a small amount of monomer. This is removed by washing with water.

The properties of nylon 6 are very similar to those of nylon 66, and the two are virtually interchangeable in their applications. The relative importance of the two materials shows marked geographical variations. In the USA and the UK, nylon 66 is the predominant material, whereas in most other regions, nylon 6 predominates.

7.3 PHENOL

Phenol is one of the oldest-established commercial organic chemicals. It was first made in the mid-nineteenth century, by extraction from coal tar, and a little is still made in this way. Manufacture by synthesis from benzene started around 1900. Production of phenol consumes about 15% of benzene.

7.3.1 Routes to phenol

It is evident that the method of choice for the manufacture of phenol from benzene would be by direct oxidation with air:

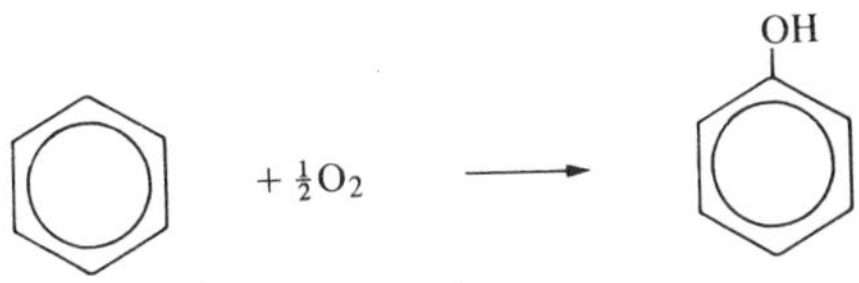

However, it has never been possible to bring about this reaction with anything like acceptable yields and conversions, so that indirect methods have to be used.

The first synthetic process used was the **sulphonation process**:

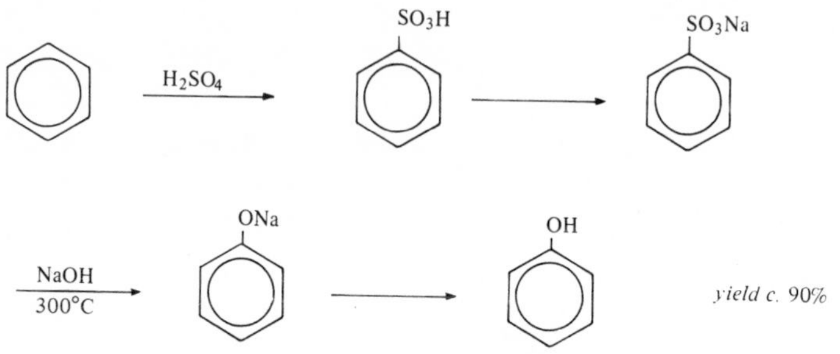

The drawbacks of this process are obvious; it has four stages, and it consumes large amounts of sulphuric acid and sodium hydroxide.

Around 1920, the **monochlorobenzene process** was introduced:

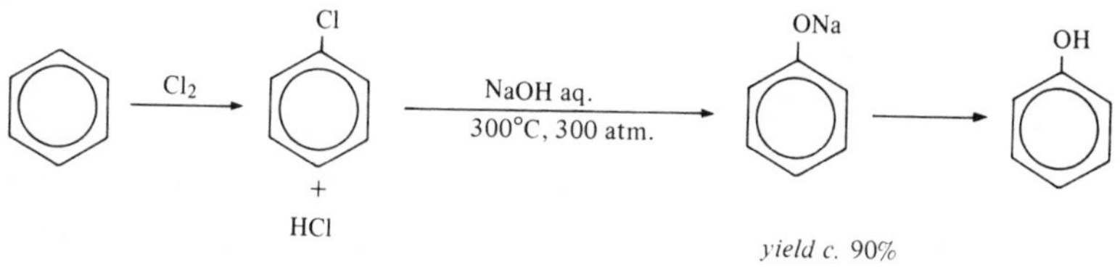

This has one less stage than the sulphonation process, but otherwise is not a great improvement. In particular, it still downgrades large amounts of ancillary raw materials.

During the 1930s, the **Raschig process** was introduced. This also goes through chlorobenzene as an intermediate, but elegantly avoids the consumption of ancillary raw materials:

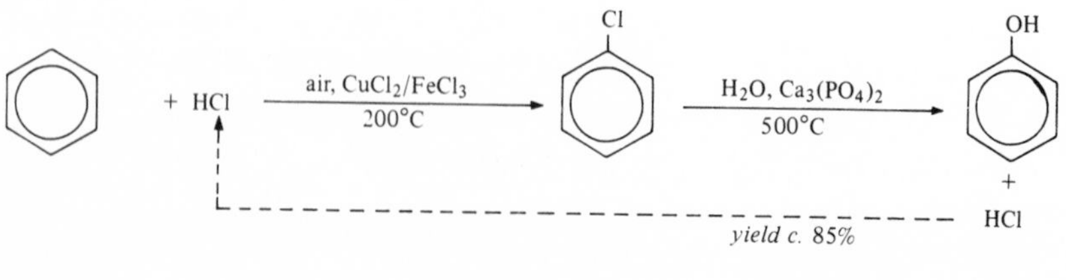

At first sight, this process would be expected to be much more economically attractive than the sulphonation and monochloro-benzene processes. It has never, however, been widely used. The reasons for this are not entirely clear. Certainly the process handles very corrosive reaction streams, and it is reputed to have suffered from corrosion problems. Both stages operate at conversions of only about 10%, so that separation and recycling costs are high. Another factor may have been that when it was first introduced, chemical engineering and control techniques may not have been adequate to cope with the difficulties of a process of this type, with two totally interlinked stages.

In the early 1950s, the *cumene process* was introduced, and this rapidly established itself as the process of choice for phenol manufacture. We shall discuss the cumene process in the next section. Two further processes were subsequently introduced, though these have not become of major importance.

The **Scientific Design process**, introduced in 1964, follows quite a different type of route from the processes we have discussed thus far:

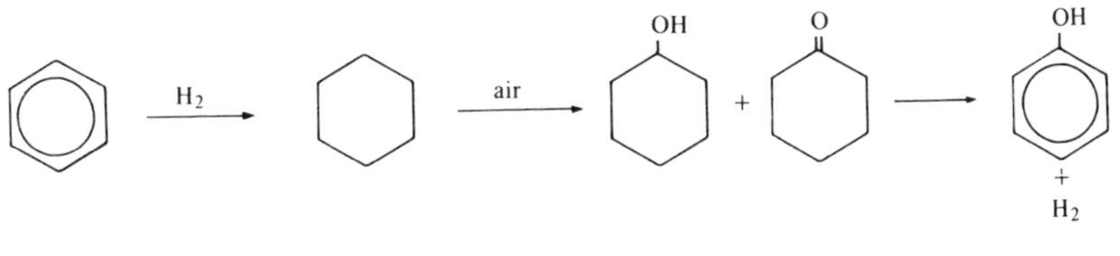

It can be seen that the first two stages of this process are based on reactions that we have already discussed. A plant to operate this process was built in Australia, but was shut down after about three years.

The **Dow toluene process** is, as its name implies, based on toluene rather than benzene, and is discussed in the next chapter.

7.3.2 The cumene process

The development of the cumene process provides an illustration of one of the ways in which chemical processes are invented. In 1944, two German chemists, Hock and Lang, who had for a number of years been carrying out research on peroxides, observed the following reaction of cumene hydroperoxide:

$$PhC(CH_3)_2COOH \xrightarrow[100°C]{dil.H_2SO_4} PhOH + CH_3COCH_3$$

Apparently not appreciating the potential commercial importance of this reaction, they published their results in one of the German chemical journals. These results were noticed by chemists in two companies, the Hercules Powder Company in the USA, and the Distillers Company in the UK. They certainly did appreciate the commercial potential of the reaction, and started to investigate it. The cumene process resulted. It was first put into commercial operation in 1953.

The process involves three stages, and produces phenol and acetone:

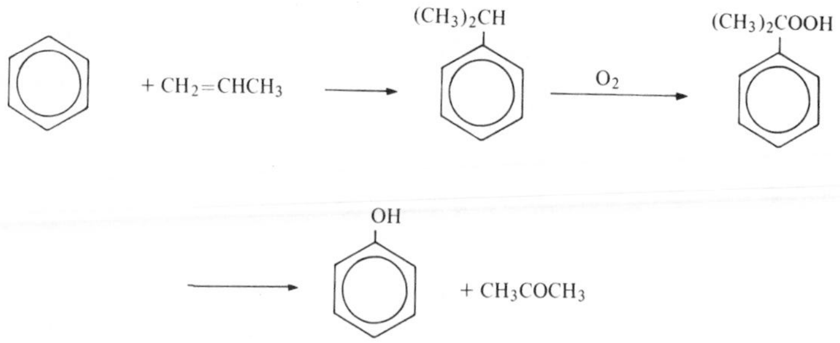

The yield of phenol is about 83%, and about 0.6 tonnes of acetone is produced for every tonne of phenol.

The first stage is usually carried out in the gas phase at about 250°C and 25 atm., over a solid phosphoric acid catalyst. An excess of benzene is used to suppress polyalkylation. The cumene is separated by distillation, and is oxidised in the liquid phase with air or oxygen at temperatures in the range 80 to 130°C. The oxidation is taken to 30% to 40% conversion, and cumene is removed from the reaction mix by vacuum distillation to give a product with a hydro-peroxide content of 75% to 85%. This is then treated with acid, e.g. dilute sulphuric acid, to decompose the hydroperoxide to phenol and acetone. The products are separated by distillation. Major by-products are acetophenone, 2-phenylpropan-2-ol, and methylstyrene.

The acid-catalysed decomposition reaction attracted a good deal of attention from academic chemists after the process was introduced, and was extensively investigated. Its mechanism is considered to be as follows:

$$
\underset{\substack{\downarrow \\ CH_3}}{\overset{\substack{CH_3 \\ \downarrow}}{PhCOOH}} \xrightarrow{H^+} \underset{\substack{\downarrow \\ CH_3}}{\overset{\substack{CH_3 \\ \downarrow}}{Ph\overset{+}{C}OOH_2}} \xrightarrow{-H_2O} \underset{\substack{\downarrow \\ CH_3}}{\overset{\substack{CH_3 \\ \downarrow}}{Ph\overset{+}{C}O}} \longrightarrow \underset{\substack{\downarrow \\ CH_3}}{\overset{\substack{CH_3 \\ \downarrow}}{PhO\overset{+}{C}}}
$$

$$
\xrightarrow{H_2O} \underset{\substack{\downarrow \\ CH_3}}{\overset{\substack{CH_3 \\ \downarrow}}{PhOC\overset{+}{O}H_2}} \xrightarrow{-H^+} \underset{\substack{\downarrow \\ CH_3}}{\overset{\substack{CH_3 \\ \downarrow}}{PhOCOH}} \longrightarrow PhOH + CH_3COCH_3
$$

The cumene process has been extremely successful, and now accounts for almost all phenol manufacture.

The economics of the process are heavily dependent on the price that can be obtained for the co-product acetone. If acetone went into gross oversupply and its price dropped to a very low level, this would no doubt stimulate interest in other ways of making phenol. However, this situation has never obtained for any length of time.

US capacity for phenol is about 1.4 million tonnes per annum.

7.3.3 Uses of phenol

Phenol has a wide range of uses. We shall discuss the three most important applications.

Phenol–formaldehyde resins

Phenol–formaldehyde resins, often called phenolic resins, were the first fully synthetic polymers made. They were introduced by Baekeland in 1910 under the trade name 'Bakelite'. Although now overshadowed by more modern polymers, they are still of substantial importance. Their manufacture is the largest single use of phenol, accounting for around 47% of consumption.

Phenol–formaldehyde resins are *thermosetting polymers*, that is to say, they set or harden by undergoing a chemical reaction during the manufacture of finished products. This contrasts with polymers such as polyethylene or poly(vinyl chloride), called *thermoplastics*, from which articles are made simply by melting the polymer, forming it into shape, and allowing it to cool.

Phenol–formaldehyde resins are made by reacting formaldehyde with phenol under either acidic or basic conditions. We will consider only the former, since this is mechanistically the most straigthforward.

In the presence of an acidic catalyst, formaldehyde is protonated:

$$CH_2=O \xrightarrow{H^+} {}^+CH_2OH$$

The carbocation thus formed rapidly attacks the highly activated *ortho*- and *para*-positions of phenol:

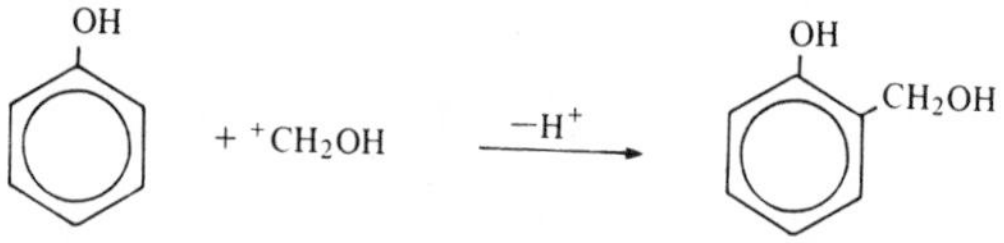

The initial products, called methylolphenols, can undergo further reactions, which result in the linking of aromatic rings by $-CH_2-$ groups ('methylene bridges'):

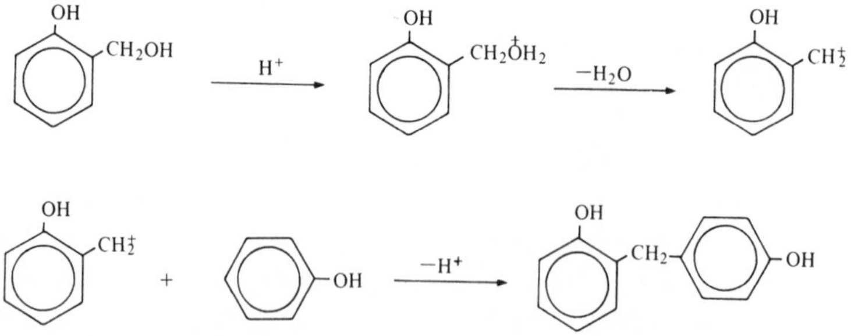

This reaction occurs at a similar rate to the initial attack by the formaldehyde, and the methylol groups react further virtually as soon as they are formed.

Typically, in making one of these resins, phenol and an aqueous solution of formaldehyde are reacted under reflux in a batch reactor, with sulphuric acid or hydrochloric acid as the catalyst. Rather less than one mole of formaldehyde per mole of phenol is used. On completion of the reaction, base is added to neutralise the catalyst, and water is distilled off, first at atmospheric pressure, and finally under vacuum. The product is a low molecular weight polymer containing structures of the type indicated in Fig. 7.1.

When the resin is used, the thermosetting process involves the reaction of further formaldehyde, added in the form of hexamethylenetetramine, with the remaining activated *ortho*- and *para*-positions, to produce a three-dimensional network polymer.

Bisphenol A

Bisphenol A is made by reacting acetone with a large excess of phenol in the presence of an acidic catalyst, e.g. hydrogen chloride, at about 50°C. The mechanism is analogous to that involved in the formation of phenol–formaldehyde resins:

$$CH_3COCH_3 \xrightarrow{H^+} (CH_3)_2\overset{+}{C}OH$$

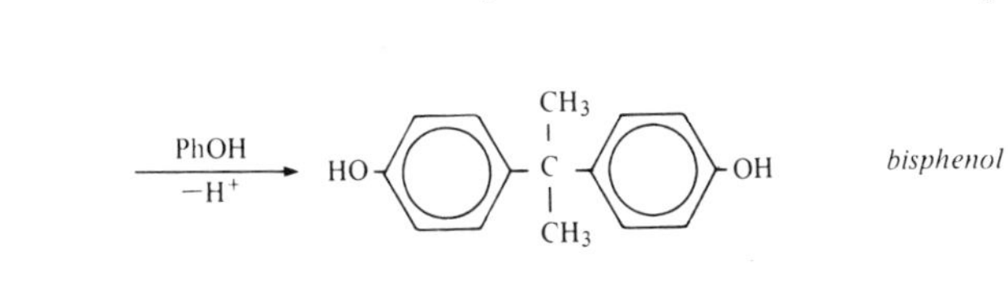

Small amounts of the 2,2'- and the 2,4'-isomers are also formed. The bisphenol A is isolated by vacuum distillation, and purified by fractional crystallisation.

The major uses of bisphenol A are in making epoxy resins and polycarbonate plastics. Its manufacture accounts for about 16% of phenol consumption.

Cyclohexanone

Hydrogenation of phenol was originally the preferred route to cyclo-

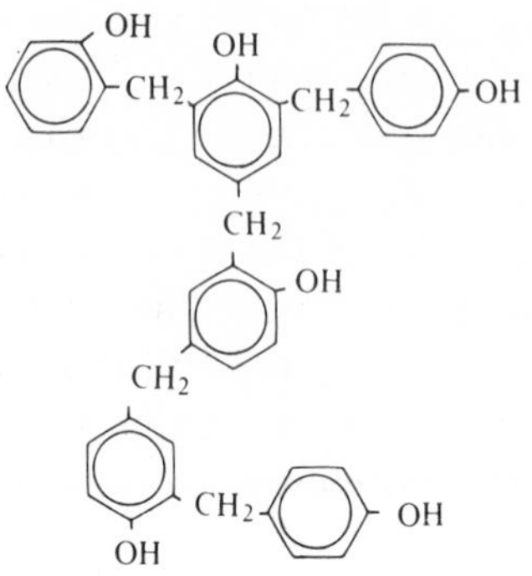

Fig. 7.1 – Typical structure present in phenol-formaldehyde resin (acid catalysed).

hexanone for use in the manufacture of caprolactam, since it was easier to make material of the required purity in this way than by the oxidation of cyclohexane. The oxidation route is now becoming increasingly important, though significant quantities of cyclohexanone are still made from phenol. About 17% of phenol is consumed in this application.

Early processes involved hydrogenation to cyclohexanol, followed by dehydrogenation of this to cyclohexanone. One-step processes are now available. In one such, phenol is hydrogenated in the liquid phase in the presence of a palladium on carbon catalyst at about 150°C and 5 atm. The yield of cyclohexanone is about 95%.

7.4 NITROBENZENE AND ANILINE

Nitrobenzene and aniline form a very old-established pair of commercial organic chemicals, having been manufactured since about 1856. Their production consumes about 5% of benzene.

Nitrobenzene is still made by essentially the same process that was originally used, except that continuous rather than batch processing is now commonly used. Benzene is nitrated using a mixture of nitric acid and sulphuric acid:

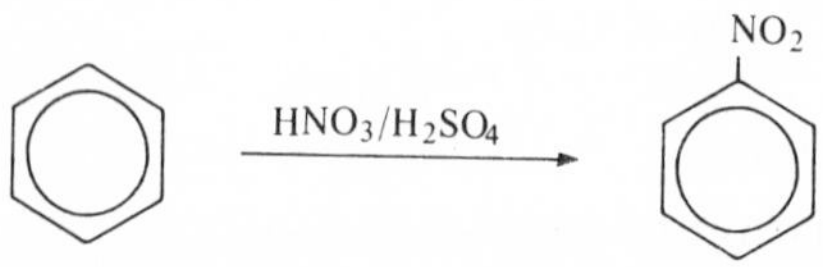

The mixture of acids, known as *nitrating mixture*, provides the nitronium ion which is the electrophilic agent in nitration:

$$HNO_3 + 2H_2SO_4 \longrightarrow NO_2^+ + H_3O^+ + 2HSO_4^-$$

The concentration of the nitronium ion depends on the ratio of the two acids, and the water content of the mix. Consequently, the 'strength' of a nitrating mixture can be adjusted to take account of the reactivity of the substrate.

In continuous nitration of benzene, the nitrating mixture typically contains 56% to 65% sulphuric acid, 20% to 26% nitric acid, and 15% to 18% water. Reaction is carried out in a stirred-flow reactor, or a series of stirred-flow reactors, at 50 to 100°C, a slight excess of benzene being used to ensure complete consumption of the nitric acid. Nitrobenzene is much less reactive towards electrophiles than benzene, and does not undergo further nitration to any significant extent under the conditions used. The nitrobenzene, which is insoluble in the spent acid, is separated, and washed with dilute sodium carbonate solution. Benzene and water are removed from the product by distillation, and the benzene is recycled. If required, the nitrobenzene may be purified by distillation, but where it is to be used for aniline manufacture, this is not necessary. The yield of nitrobenzene is very high, around 96% to 99%. The spent acid, which is essentially diluted sulphuric acid, is reconcentrated and recycled.

About 98% of the nitrobenzene made is converted into aniline. The original method of doing this was by reduction with iron powder and dilute hydrochloric acid, but this has been entirely supplanted by catalytic hydrogenation:

$$PhNO_2 + 3H_2 \longrightarrow PhNH_2 + 2H_2O$$

Both liquid phase and gas phase processing are used, but the latter is more important. In one gas phase process, the reaction is carried out at 270 to 290°C and 1 to 5 atm. in a fluidised bed of copper catalyst, using a hydrogen to nitrobenzene ratio of about $9:1$. Yields of 98% or more are quoted.

US capacity for aniline is about 640 000 tonnes per annum.

Aniline has a wide variety of uses, mostly relatively small scale. The only one carried out on a scale large enough to be considered a petrochemical process is the manufacture of 4,4′-diaminodiphenylmethane, which is then converted into 4,4′-diphenylmethane di-iso-cyanate (MDI):

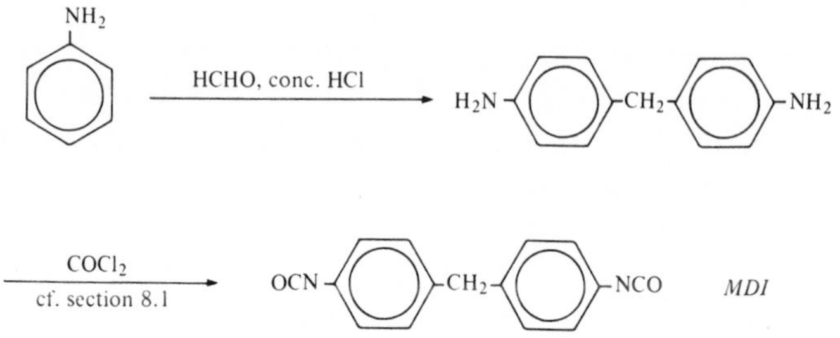

MDI (US capacity about 380 000 tonnes per annum) is used for making rigid polyurethane foams and other polyurethane products.

7.5 MALEIC ANHYDRIDE

Manufacture of maleic anhydride began in about 1930. The route used then, gas phase catalytic oxidation of benzene, still accounts for most manufacture today:

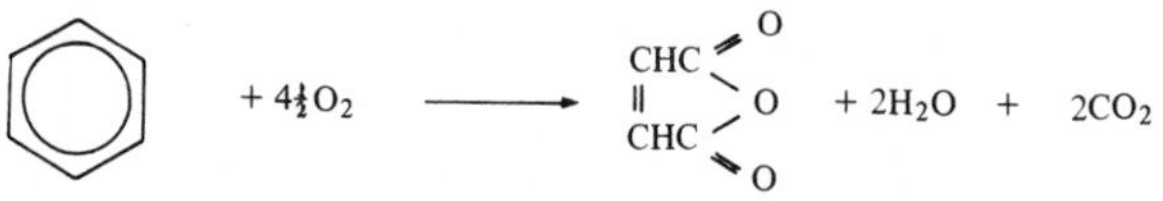

The oxidation is carried out with air, in large excess, so that the mixture is outside the lower explosive limit, over a vanadium pen-toxide/molybdenum oxide catalyst at 350 to 400°C and atmospheric pressure. The exit gases are cooled to below 200°C and then pass to a condenser where about 50% of the maleic anhydride condenses. The rest is absorbed in water, where it hydrolyses to maleic acid. The conversion of benzene is high, around 95%, but yields are only moderate at about 60% to 70%. It can be seen that this process is intrinsically wasteful of raw materials, since two of the six carbon atoms in benzene are converted into carbon dioxide. Processes based on the oxidation of n-butane and n-butenes are potentially more efficient in this respect, and also have the advantage of using cheaper hydrocarbons. Such processes are used to some extent, and may become more important in the future. At present, however, the bulk of maleic anhydride is made from benzene. Total US capacity for maleic anhydride is about 180 000 tonnes per annum.

The major use of maleic anhydride, consuming about 55% of the total, is in making unsaturated polyester resins for use in the manufacture of glass fibre reinforced plastics, and in surface coatings. These resins are made by reacting a mixture of maleic anhydride and phthalic anhydride with propylene glycol (section 8.4.1).

7.6 DETERGENT ALKYLATE AND ALKYLBENZENE-SULPHONATE DETERGENTS

Sodium alkylbenzenesulphonates are the most important of the surface-active agents in domestic use. They are the active components of most heavy-duty domestic detergent powders, e.g. Daz, Omo, Surf, Tide, etc. Optimum detergent properties are obtained when the alkyl group has about twelve carbon atoms.

Until the 1960s, detergent alkylate was usually made by alkylating benzene with propylene tetramer (section 4.6). This gives a product with highly branched side chains, and the detergent derived from this is 'biologically hard', that is, it is resistant to degradation by micro-organisms. When the use of synthetic detergents became widespread in the 1950s, very major problems of foam formation in sewage works and on rivers developed, and it became clear that this was due to residues of detergents in the waters involved. The problem was solved, after a great deal of investigation, by changing over to the use of the so-called linear alkylbenzenesulphonates (LABS), which are biodegradable. The description 'linear' is not strictly correct, as we shall see.

The detergent alkylate used for making LABS is made by alkylating benzene with straight chain alkenes, which may be terminal or non-terminal, or with straight chain chloroalkanes. Depending on the source, the alkenes used may be essentially dodecene, or may be mixtures spanning the range C_{10} to C_{14}. Manufacture of the alkylating agents is discussed in sections 3.10, 10.4 and 10.5.

Alkylation using alkenes is carried out in the presence of hydrogen fluoride, at temperatures in the range 5 to 70°C, a large excess of benzene being used to suppress polyalkylation. The product, in about 90% yield, is isolated by vacuum distillation. It is evident that when non-terminal alkenes are used, the alkyl group attached to the ring will be of the form $-CHR_1R_2$, e.g.:

$$C_6H_{13}CH{=}CHC_4H_9 \longrightarrow C_6H_{13}\overset{+}{C}HCH_2C_4H_9 \xrightarrow{\ C_6H_6\ }$$

It might be supposed that when 1-alkenes are used, the group attached has the structure −CHRMe. However, the carbocations initially formed undergo some rearragement by 1,2-hydride shifts, so that a substantial proportion of the alkyl groups have the same type of structure as those introduced by alkylation with non-terminal linear alkenes. It is clearly important that no significant amount of rearrangement of the carbon skeleton of the carbocation occurs, since this would give rise to biologically hard material.

Alkylation with chloroalkanes is carried out in the presence of aluminium chloride at temperatures in the range 40 to 85°C. The yield is rather lower than in the alkene-based process, being around 80%. This lower yield is, to some extent at least, counterbalanced by the fact that the alkene-based process involves the handling of large amounts of hydrogen fluoride, a rather hazardous material.

The alkylbenzenes are sulphonated using either 20% oleum or sulphur trioxide. The sulphonic acid group mainly enters the *para* position.

When oleum is used, the reaction is carried out at about 50°C, and then water is added to dilute the remaining acid to a concentration of about 70% to 80%. Two phases form, a spent acid phase, and a product phase which also contains about 7% to 10% of sulphuric acid. This is separated and neutralised with either sodium carbonate or sodium hydroxide, to give paste of sodium alkylbenzenesulphonate containing some sodium sulphate.

Sulphonation with sulphur trioxide involves contacting the alkylate with a dilute mixture of sulphur trioxide in air. The reaction is extremely vigorous, and special reactor designs have to be used to achieve good gas–liquid contact and heat transfer. The advantage of this type of process is that the product obtained is virtually free from sodium sulphate, and no spent acid is produced.

US capacity for linear alkylbenzenes is about 273 000 tonnes per annum.

PROBLEMS AND EXERCISES

1. Since naphtha contains cyclohexane, and this is converted into

benzene during reforming, the production of cyclohexane by hydrogenation of benzene seems rather a roundabout route. Suggest why it is used.

2. Cyclohexane plants are often situated next to catalytic reformers. Proximity to a supply of benzene is one reason, but there is another, which is in fact more important. What is it?

3. Suggest reaction paths for the formation of adipic acid and hexanoic acid from cyclohexyl hydroperoxide during the oxidation of cyclohexane.

4. Look up the mechanism of the Beckmann rearrangement in a textbook of organic chemistry (unless you already know it!) and then show how it operates in the formation of caprolactam.

5. Acetophenone and 2-phenyl-2-propanol are by-products in the air oxidation stage of the cumene process. Suggest mechanisms for their formation.

6. What product would you expect from the reaction of 2,4-dimethylphenol and formaldehyde in the molar proportions 2:1 under conditions similar to those used in making phenol—formaldehyde resins?

8

Toluene and xylene derivatives

The range of chemical uses of toluene and xylenes is much more limited than that of benzene. o-Xylene and p-xylene have one large-scale use each, in the manufacture of phthalic anhydride and terephthalic acid respectively. In the case of toluene, apart from benzene, discussed in section 6.5, tolylene di-isocyanate is the only derivative which is widely and regularly made on a fairly large scale. Benzoic acid is made on a large scale at a very limited number of locations, and trinitrotoluene is made on a large scale intermittently, when required for military operations. Toluene has a number of small tonnage uses, e.g. in the manufacture of saccharin, but these are outside the scope of this book.

8.1 TOLYLENE DI-ISOCYANATE

Tolylene di-isocyanate, generally called TDI, is used in making polyurethanes, notably flexible polyurethane foams, materials which have a wide range of uses in car upholstery, furniture, bedding, carpets, textiles, and other outlets. Polyurethanes were developed in Germany during World War II, and TDI was manufactured there on a small scale at that time. Major development of polyurethanes occurred during the 1950s, and the manufacture of TDI on a substantial scale dates from that time. Current US capacity is about 310 000 tonnes per annum.

The process used for TDI manufacture is an unusually complicated and difficult one for large-scale operation, and this has tended to limit the number of companies that have become involved in TDI production. The route used is as follows:

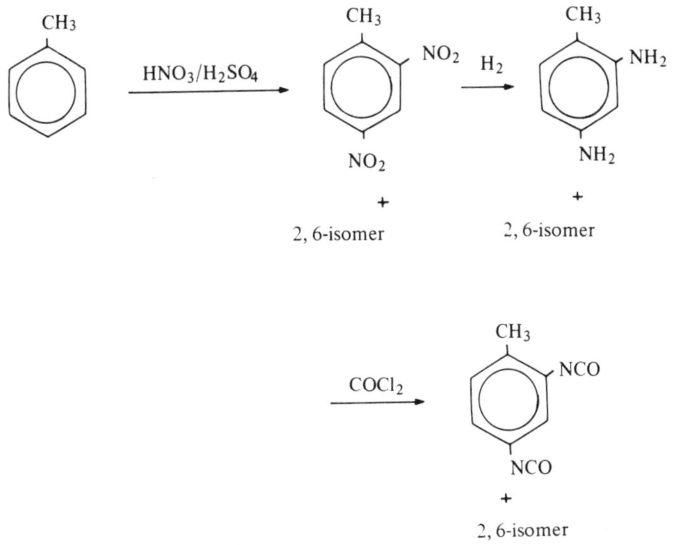

The main commercial product is a mixture of 80% 2,4-tolylene di-isocyanate and 20% 2,6-tolylene di-isocyanate, this being the ratio of isomers obtained by dinitration of toluene. Much smaller quantities of a 65—35 mixture and of pure 2,4-tolylene di-isocyanate are also used. Their production involves separation of either the di-isocyanates or of o- and p-nitrotoluenes, and they are consequently more expensive than the 80—20 mixture.

The nitration is carried out in two stages. Typically, toluene is mononitrated by treatment with a nitrating mixture of composition around 19% nitric acid, 60% sulphuric acid, and 21% water, at about 30 to 45°C, and then the mononitrotoluenes are reacted further under more forcing conditions, e.g. with a mixture of about 35% nitric acid and 65% sulphuric acid at 65 to 80°C.

Hydrogenation of the dinitrotoluenes is carried out in the liquid phase in the presence of nickel, platinum, or palladium catalysts. For example, in one process hydrogenation is at 90°C and 50 atm. in solution in methanol, with a Raney nickel catalyst in suspension.

The reaction of the tolylenediamines with carbonyl chloride is carried out in the liquid phase in dilute solution in an inert solvent such as chlorobenzene or o-dichlorobenzene. Typically, a series of stirred flow reactors is used, with the reaction temperature increasing from about 20 to 190°C. The TDI is purified by vacuum distillation. Yields from 75% to 88%, based on toluene, are quoted.

Overall, this process is far removed from being an ideal process for large scale operation. The chemical route followed is complicated and inelegant, and each of the three stages involves some quite severe engineering problems. Notable amongst these is that all stages handle hazardous materials. Thus, nitrotoluene is toxic and potentially explosive, hydrogen is highly flammable, and tolylenediamine, carbonyl chloride and TDI are all toxic. The process gives four moles of hydrogen chloride for every mole of TDI, so that the return, if any, obtained for this clearly has a significant effect on the economics of the process.

Quite a lot of effort has been put into a quest for better routes to TDI, and processes have been developed in which it is produced direct from the dinitrotoluenes, e.g.

$$C_7H_6(NO_2)_2 + 6CO + 2EtOH$$

$$\xrightarrow{\text{catalyst}} C_7H_6(NHCO_2Et)_2 + 4CO_2$$

$$C_7H_6(NHCO_2Et)_2 \xrightarrow{\text{pyrolysis}} C_7H_6(NCO)_2 + 2EtOH$$

At the time of writing, it does not appear that any process of this type is operated commercially.

Polyurethanes are made by reacting TDI, or other di-isocyanates, with polyhydroxy compounds, e.g. polyether polyols made from propylene oxide (section 4.3.1). The reaction which gives rise to the urethane linkage is as follows:

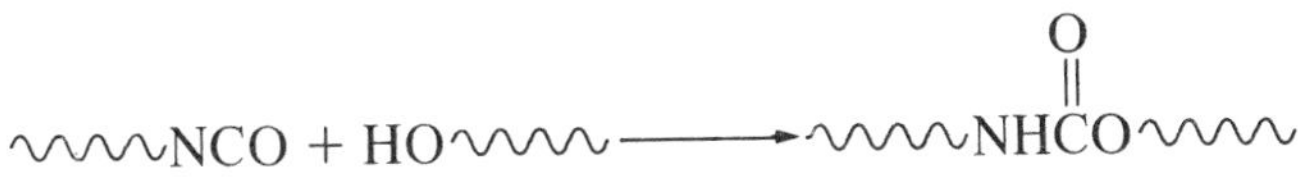

$$\text{\Large \textasciitilde\textasciitilde\textasciitilde NCO} + \text{HO}\text{\textasciitilde\textasciitilde\textasciitilde} \longrightarrow \text{\textasciitilde\textasciitilde\textasciitilde NHCO}\text{\textasciitilde\textasciitilde\textasciitilde}$$

$$\overset{O}{\underset{\|}{}}$$

However, there are a number of other reactions which occur during the formation of polyurethane products, and the chemistry and technology involved is very complex.

8.2 BENZOIC ACID

Benzoic acid is made by oxidation of toluene in the liquid phase with air:

$$\text{PhCH}_3 + 1\tfrac{1}{2}O_2 \longrightarrow \text{PhCO}_2\text{H} + H_2O$$

Typically, the conditions are similar to those used for oxidising cyclohexane to cyclohexanol and cyclohexanone. Thus, a soluble cobalt salt is used as catalyst, and the temperature and pressure are in the range 140 to 160°C and 4 to 10 atm. The initial product is a hydroperoxide, formed by reactions analogous to those we have discussed previously for this type of oxidation. This cleaves across the oxygen–oxygen bond, and further reactions of the radical thus formed give rise to benzoic acid:

$$
\text{PhCH}_2\text{OOH} \longrightarrow \text{PhCH}_2\text{O·}
\begin{cases}
\xrightarrow{-[\text{H·}]} \text{PhCHO} \xrightarrow{\text{oxdn.}} \text{PhCO}_2\text{H} \\
\xrightarrow{+[\text{H·}]} \text{PhCH}_2\text{OH} \xrightarrow{\text{oxdn.}} \text{PhCO}_2\text{H}
\end{cases}
$$

The process is operated at about 35% conversion and gives a yield of about 90%.

Oxidation in acetic acid solution in the presence of bromide ions (the Amoco, process, cf. section 8.5) is also used.

In general, benzoic acid has only small-scale applications, and most plants operate at a scale of a few thousand tonnes per annum. However, it is used to a limited extent as an intermediate in processes for making caprolactam and phenol from toluene, and where this is done it is made on a typical 'petrochemical' scale.

Manufacture of caprolactam from toluene is carried out by the process outlined below:

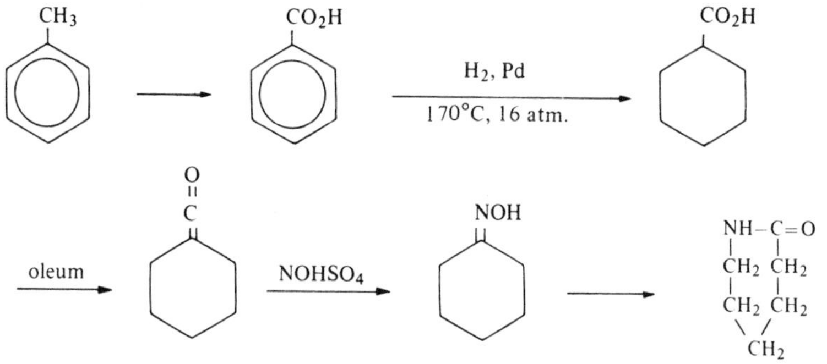

This process was developed by the Italian firm Snia Viscosa, and is operated by them in Italy. It is also reputed to be operated in the USSR. As we have already noted in section 7.2.6, most caprolactam is made from cyclohexanone, derived from benzene.

Phenol may be made from toluene by a process developed by Dow during the late 1950s. In this, molten benzoic acid containing a copper salt catalyst is oxidised with air to give phenol and carbon dioxide:

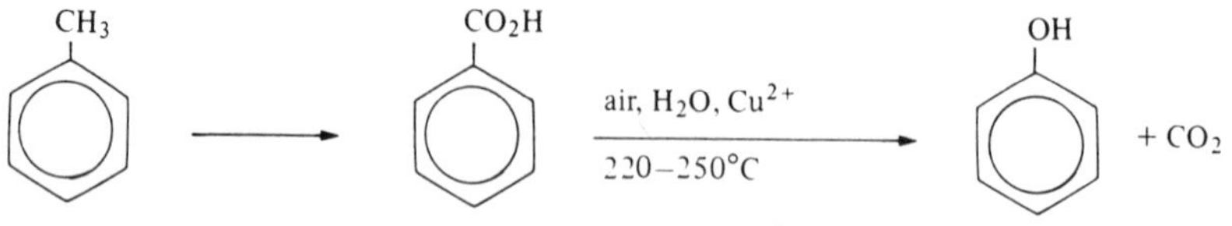

The yield of phenol, based on toluene, is quoted at around 73%. This process has been operated in the USA and Holland, but no plants have been built for a number of years, and it would appear that the process is not competitive with the cumene process.

8.3 TRINITROTOLUENE

Trinitrotoluene (TNT) was first used as a military explosive during World War I. It rapidly became the most important of the military explosives, and has remained so ever since. Although it has some civil applications, these use relatively small quantities. Consequently, the scale of manufacture of TNT varies greatly from peacetime to wartime. Under war conditions, it provides a major demand for toluene.

In outline, the process used for making TNT is simple:

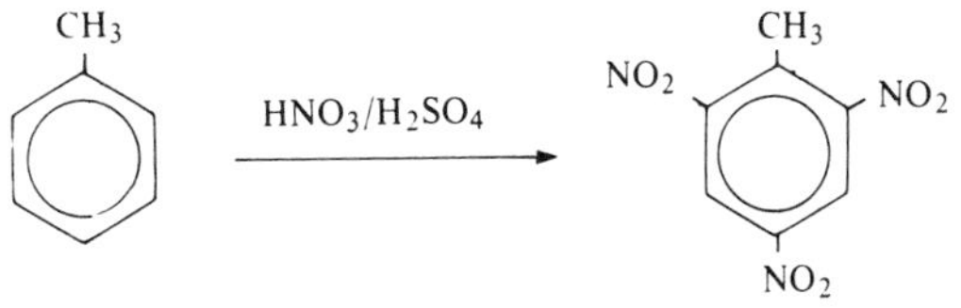

However, there is a great deal of variation between processes in fine detail. For example, some processes involve three reaction stages utilising successively more vigorous conditions; other carry out the nitration in one or two stages.

After nitration, the crude TNT is separated from the spent acid, in which it is insoluble, and washed acid free with water. The yield is 85% to 90%.

8.4 PHTHALIC ANHYDRIDE

Phthalic anhydride first became commercially important during the nineteenth century as an intermediate for dyestuffs manufacture. Early methods of production involved oxidising naphthalene with agents such as nitric acid or sulphuric acid. In 1918, manufacture by air oxidation of naphthalene in the gas phase was started:

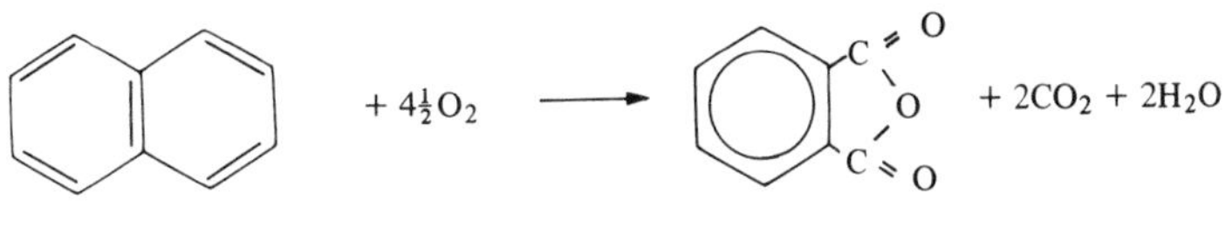

This became the sole method of manufacture for almost thirty years.

Naphthalene is a by-product of coal carbonisation, and the amounts available from this source are limited by the amounts of coal being carbonised. As demand for phthalic anhydride developed, it became clear that supplies of naphthalene would become inadequate, and an alternative raw material was sought. o-Xylene is an obvious candidate, and it was first used for phthalic anhydride manufacture in 1946, in the USA. During the 1960s its use became widespread, and it is now the dominant raw material, probably accounting for 80% to 90% of production.

The process used for the production of phthalic anhydride from o-xylene is broadly similar to that based on naphthalene. It involves catalytic oxidation with air, in the gas phase:

The catalysts used are vanadium pentoxide supported on silica or silicon carbide and promoted with various other metal oxides, e.g. titanium dioxide. The reaction is strongly exothermic, and the necessity for very good heat transfer to control the temperature at the required level in the range 375 to 410°C has a dominant effect on the technology. Typically, tubular reactors cooled by circulating molten salt mixture are used. A large excess of air is used, so that the reaction mixture is outside the lower explosive limit.

The gas stream from the reactor passes to 'switch condensers' in which it is contacted with cooled finned tubes. The phthalic anhydride condenses and solidifies on the tubes, and then the gas stream is switched to another condenser while the solidified product is melted off the tubes and collected. It is held at an elevated temperature for a period, to dehydrate the small amount of phthalic acid that is present, and to convert various by-products into high boiling materials, and then it is purified by vacuum distillation. Yields of up to 78% are claimed.

US capacity for phthalic anhydride is about 550 000 tonnes per annum.

8.4.1 Uses of phthalic anhydride

Phthalic anhydride has three major uses which between them consume around 95% of production. The remainder is used in making a variety of products, including dyestuffs.

The manufacture of **dialkyl phthalates** is the most important single outlet for phthalic anhydride, accounting for about 50% of consumption in the USA, and more than 60% of consumption in Western Europe and Japan. Dialkyl phthalates are used as *plasticisers* for polymers, notably for poly(vinyl chloride). Unplasticised poly(vinyl chloride) is a hard, rigid polymer, familiar as guttering and drain pipes on houses. The addition of a plasticiser, in proportions up to about 45%, converts it into a soft and flexible material. About 80% of total dialkyl phthalate use is in poly(vinyl chloride).

Phthalate plasticisers are made from a variety of alcohols containing up to 13 carbon atoms. However, the most important are those made from C_8 alcohols; for most purposes these have the optimum combination of properties. The Oxo process (section 4.5) provides the source of most plasticiser alcohols.

Dialkyl phthalates are prepared by reacting the alcohol and phthalic anhydride in the presence of an acidic catalyst, usually sulphuric acid, at temperatures around 150 to 200°C:

The product is purified by vacuum distillation. Both batch and continuous processing are used.

Unsaturated polyesters are made by the reaction of phthalic anhydride and maleic anhydride with propylene glycol. They are low molecular weight polymers containing the following types of structure:

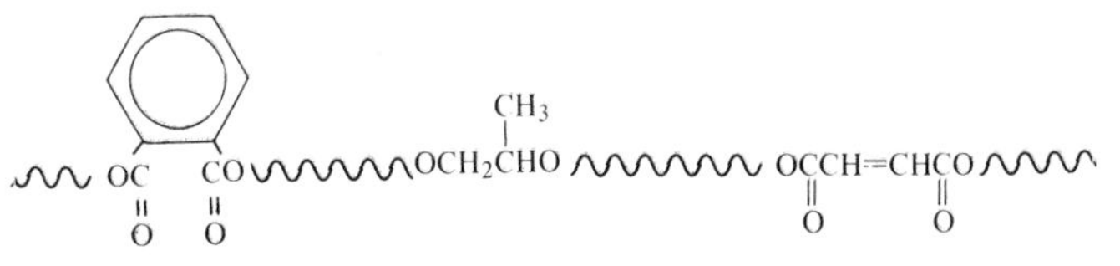

The carbon−carbon double bonds in the maleate groups provide active centres through which these polymers can be crosslinked. When they are used, for example in making glass fibre reinforced plastics, they are mixed with styrene and a free radical initiator, and the styrene copolymerises with the polyester to give a cross-linked polymer.

Alkyd resins consume roughly the same proportion of phthalic anhydride production as the unsaturated polyesters. They are low molecular weight polyesters with complex, ill-defined structures, made by reacting phthalic anhydride with glycerol, vegetable oils and/or fatty acids, and various other components. They are mainly used in surface coatings, e.g. in oil-based paints.

8.5 TEREPHTHALIC ACID AND DIMETHYL TEREPHTHALATE

Manufacture of terephthalic acid and dimethyl terephthalate is the only major outlet for p-xylene. These materials are required for the production of poly(ethylene terephthalate) for use as fibre, and, to a lesser extent, as a plastic.

Poly(ethylene terephthalate) fibres were invented in 1941, and were first made commercially around 1950. Terephthalic acid had not previously been made on a large scale.

The reader will by this stage no doubt appreciate that one potentially attractive route to terephthalic acid is the liquid phase free radical oxidation of p-xylene. However, in the early days of poly(ethylene terephthalate) manufacture there were two problems associated with this. Firstly, p-xylene had not previously been produced commercially, and was expensive. Secondly, the oxidation of p-xylene with air under conventional liquid phase oxidation conditions does not yield terephthalic acid, but rather p-toluic acid:

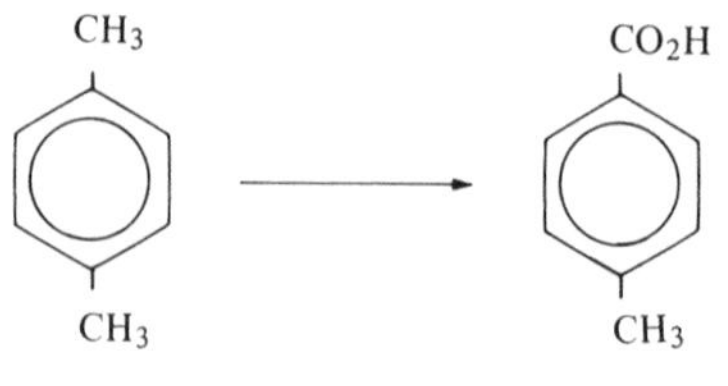

Because of these and other factors, a variety of processes were developed, some based on p-xylene, and some on other hydrocarbons.

The earliest process based on *p*-xylene involved oxidation by nitric acid:

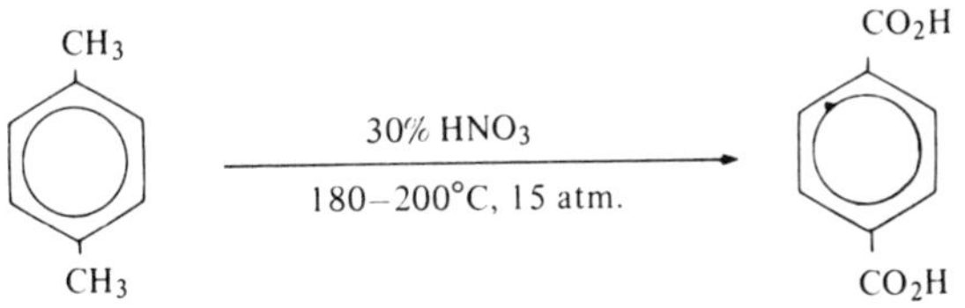

CH$_3$ — [ring] — CH$_3$ $\xrightarrow[\text{180–200°C, 15 atm.}]{\text{30\% HNO}_3}$ CO$_2$H — [ring] — CO$_2$H

The product obtained by this method contains substantial amounts of various impurities. Since terephthalic acid is very difficult to purify, being involatile and having a low solubility in common solvents except under extreme conditions, it was converted to dimethyl terephthalate, which was then purified by vacuum distillation and/or recrystallisation. As we shall see, the early poly(ethylene terephthalate) processes were based on dimethyl terephthalate rather than the free acid.

Although *p*-toluic acid is very resistant to air oxidation, its methyl ester is fairly readily oxidised, so that dimethyl terephthalate can be made from *p*-xylene by a two-stage oxidation combined with esterification. During the 1950s and 1960s a number of processes of this type were developed, and became the main source of dimethyl terephthalate. The most important of these, the *Witten process*, operates as follows:

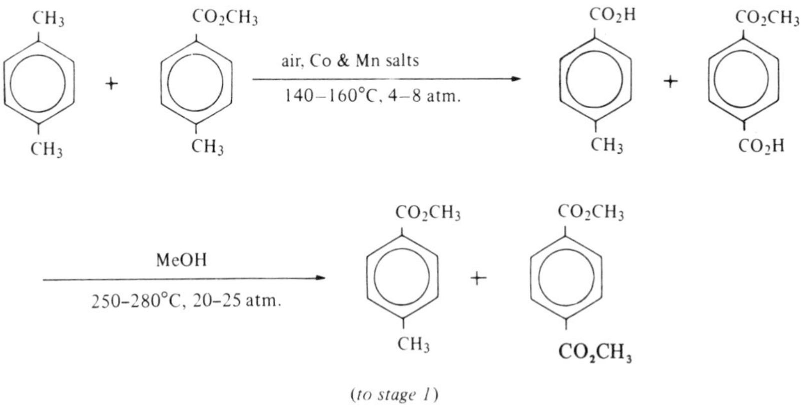

CH$_3$—[ring]—CH$_3$ + CO$_2$CH$_3$—[ring]—CH$_3$ $\xrightarrow[\text{140–160°C, 4–8 atm.}]{\text{air, Co \& Mn salts}}$ CO$_2$H—[ring]—CH$_3$ + CO$_2$CH$_3$—[ring]—CO$_2$H

$\xrightarrow[\text{250–280°C, 20–25 atm.}]{\text{MeOH}}$ CO$_2$CH$_3$—[ring]—CH$_3$ + CO$_2$CH$_3$—[ring]—CO$_2$CH$_3$

(*to stage 1*)

The situation with regard to the manufacture of terephthalic acid was transformed during the 1960s by the introduction of the **Amoco process**. In this, p-xylene is oxidised by air, in the liquid phase in solution in acetic acid at about 200°C and under 20 atm. pressure, with a catalyst system containing cobalt, manganese and bromide ions in solution. Terephthalic acid is formed in 90% to 95% yield, with almost complete conversion of the p-xylene. The reaction mixture is extremely corrosive, and titanium-lined reactors have to be used.

The terephthalic acid produced contains only small amounts of impurities, mainly p-toluic acid and p-carboxybenzaldehyde, and these can be removed by relatively simple procedures, to give poly-merisation grade acid. Thus, the acid is dissolved in water at about 250°C and 40 to 50 atm., and the solution is treated with hydrogen in the presence of a hydrogenation catalyst such as palladium on charcoal. This hydrogenates the aldehyde. The solution is then cooled to 100°C, and pure terephthalic acid crystallises out.

Since the Amoco process was introduced, there has been a trend towards the use of terephthalic acid rather than dimethyl terephthalate in polymerisation processes (see next section).

Bromine plays a key role in the mechanism of the oxidation in the Amoco process. We have already seen that under conventional conditions, liquid phase air oxidation of p-xylene gives p-toluic acid. This does not react further to any significant extent because the hydrogen atoms of its methyl group are resistant to attack by the radicals present. They are readily attacked, however, by bromine atoms:

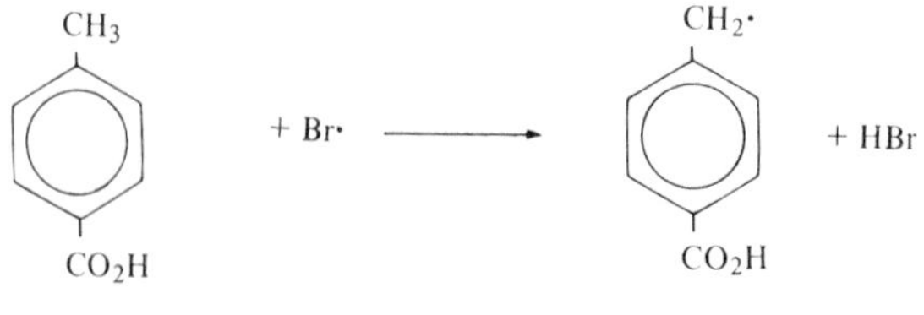

Once hydrogen atom abstraction has occurred, oxidation will follow the type of sequence we have already met:

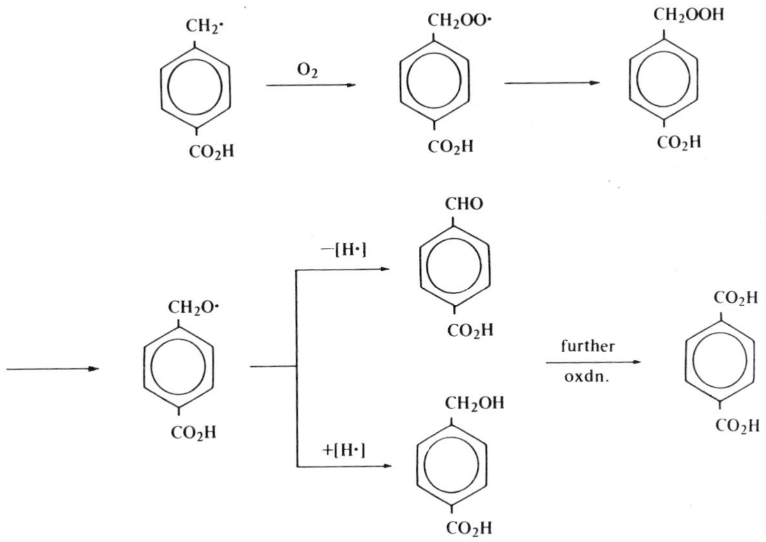

Bromine atoms are regenerated from the hydrogen bromide by hydrogen atom abstraction:

$$HBr + Rad\cdot \longrightarrow Br\cdot + RadH$$

Initially, they may be assumed to be produced by oxidation of bromide ions by one of a variety of species:

$$Br^- \xrightarrow{-[e^-]} Br\cdot$$

A number of other processes have since been developed for the air oxidation of p-xylene. One uses oxidation in acetic acid solution in the presence of a very high concentration of cobalt acetate catalyst. Others involve co-oxidation of p-xylene with a readily oxidised material such as acetaldehyde. Advantages claimed over the Amoco process are the avoidance of the need for titanium-lined reactors, and a product that is more easily purified. However, the Amoco process still accounts for the bulk of the world's terephthalic acid production.

US capacities for dimethyl terephthalate and terephthalic acid respectively are about 1.6 million tonnes per annum and 1.2 million tonnes per annum.

8.5.1 Poly(ethylene terephthalate)

The original method of manufacture of poly(ethylene terephthalate) was based on dimethyl terephthalate. This was partly because of the greater ease of production of the ester, already discussed, and partly because it is somewhat easier to carry out the polymerisation with the ester than with the acid. However, the use of the ester carries a penalty in the costs involved in using and recycling methanol, and since polymerisation-grade terephthalic acid became readily available there has been a steady shift towards its use. At present, about 40% of poly(ethylene terephthalate) is made direct from the acid. With both starting materials, the polymerisation is carried out in two stages.

In the first stage of an acid-based process, terephthalic acid and a 10% to 50% molar excess of ethylene glycol are reacted at 240 to 260°C and about 4 atm., water being allowed to escape as vapour as the reaction proceeds. The product is bis(2-hydroxyethyl) terephthalate and low molecular weight oligomers:

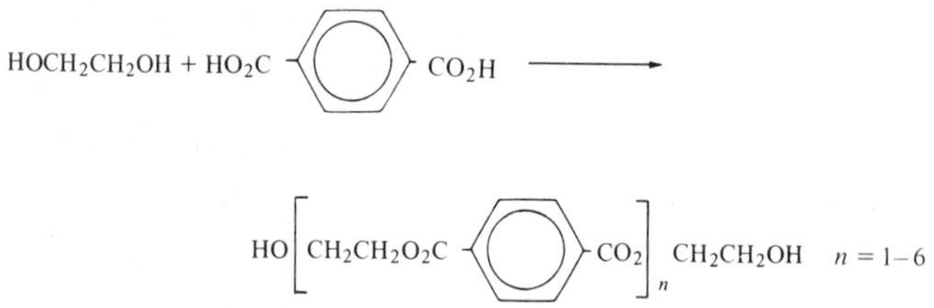

A catalyst for transesterification, e.g. antimony trioxide, is then added, and the reaction is continued. Polymerisation occurs by transesterification, with elimination of ethylene glycol:

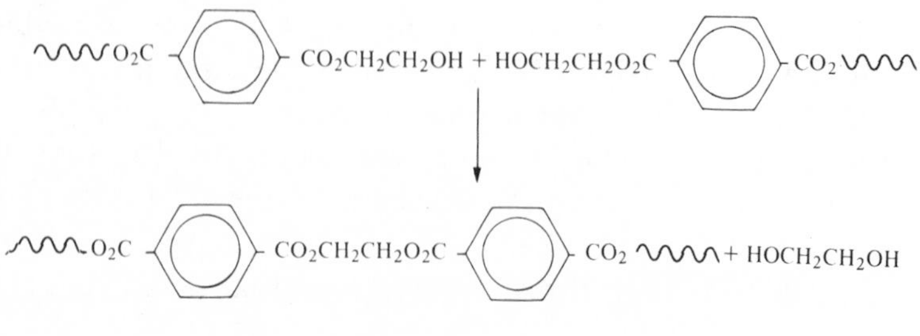

The glycol is distilled off as it is formed, and the temperature is raised by stages to 280 to 290°C, while the pressure is reduced to about 1 mm Hg.

When dimethyl terephthalate is used, an ester interchange catalyst, usually manganese acetate, is added in the first stage, and a rather larger excess of glycol is used (molar ratio glycol to acid about 2.4:1). Methanol is distilled off as the reaction proceeds. The second stage of the polymerisation is carried out in the same way as when manufacture is based on the free acid.

Originally, poly(ethylene terephthalate) was made by batch processing, but continuous processing is now also widely used.

At first sight, it is difficult to see why poly(ethylene terephthalate) is not made in a single stage, simply by reacting equimolar quantities of ethylene glycol and terephthalic acid. Let us consider this.

In this type of polymerisation, called stepwise or condensation polymerisation, high molecular weights can only be achieved if there is a very close balance between the functional groups present. If one of the groups is present in excess, at a certain stage of the reaction, all molecules will have this group at each end, and reaction will necessarily stop. Consequently, very careful proportioning of monomers is normally of extreme importance in stepwise polymerisation. In polyesterifications, even with exact proportioning of monomers at the start of the reaction, problems still arise, since side reactions which result in loss of hydroxy groups occur, and it is difficult to make high molecular weight polymers by simple polyesterification. Polymerisation by transesterification provides an elegant way round this difficulty.

PROBLEMS AND EXERCISES

1. Why are more forcing conditions required for the nitration of nitrotoluene than of toluene?
2. Write down a balanced equation for the reaction of tolylenediamine with carbonyl chloride.
3. What do you think is the main advantage of the Amoco process over the 'conventional' process for benzoic acid manufacture? Can you think of a disadvantage?
4. Why is dimethyl terephthalate more volatile and more readily soluble than terephthalic acid?
5. Draw an outline flow scheme for the Witten process for dimethyl terephthalate.

6. Why is it necessary to use elevated pressures when recrystallising terephthalic acid from water?
7. Suggest side reactions which might lead to the loss of hydroxy groups during the production of poly (ethylene terephthalate).

9

Steam reforming
and related processes

Steam reforming is the third of the basic building block petrochemical processes. It is operated to produce mixtures of nitrogen and hydrogen for ammonia production, and mixtures of carbon monoxide and hydrogen for the production of methanol and for a variety of other purposes. It is one of the quirks of the technical jargon of the chemical industry that the term *synthesis gas* is used to describe both mixtures of nitrogen and hydrogen and of carbon monoxide and hydrogen. Fortunately, it is usually clear from the context which is meant.

Where natural gas is available, this is the feed normally used for steam reforming. Dry natural gas is used as such. Wet natural gas is usually first stripped of its ethane, propane, and butane. Where no natural gas is available, naphtha is used.

9.1 THE REFORMING REACTIONS

With natural gas as feedstock, the basic reforming reaction is:

$$CH_4 + H_2O \rightleftharpoons CO + 3H_2$$

A variety of materials act as catalysts for the reaction, but supported nickel catalysts are invariably used in commercial processes. The reaction is reversible and endothermic.

There are a number of reactions which can result in carbon formation during reforming:

$$CO + H_2 \rightleftharpoons C + H_2O$$

$$2CO \rightleftharpoons C + CO_2$$

$$CH_4 \rightleftharpoons C + 2H_2$$

It is essential that the formation of carbon is avoided, since it would deposit on the catalyst and inactivate it. Use of an excess of steam helps prevent this, by causing the first of the above reactions to run from right to left.

The *shift reaction* also occurs under reforming conditions:

$$CO + H_2O \rightleftharpoons CO_2 + H_2$$

This is exothermic, and at reforming temperatures the equilibrium constant is small, so that only a little carbon dioxide is formed. As we will see later, the shift reaction provides a means of removing carbon monoxide from the gas stream when this is necessary.

When naphtha is reformed the reactions involved are clearly more complicated than than those in methane reforming, and they are not in fact very well understood. Overall, the individual components of naphtha rapidly react with steam to give carbon monoxide and hydrogen; since naphtha typically has a hydrogen to carbon ratio of about 2.1:1, it is convenient to represent the main reforming reaction as follows:

$$CH_{2.1} + H_2O \rightleftharpoons CO + 2.05H_2$$

As in methane reforming, a certain amount of carbon dioxide is formed by the shift reaction. Also, a small amount of methane is formed by the reverse of the methane-reforming reaction.

The tendency towards carbon formation is greater than with methane, and the catalysts used for naphtha reforming are specially designed to counteract this. Typically, in addition to nickel, they contain potassium oxide, one effect of which is to catalyse the reaction of carbon with steam.

9.2 STEAM REFORMING FOR AMMONIA

The manufacture of synthetic ammonia is one of the most important operations carried out by the chemical industry. The bulk of ammonia is used for fertiliser manufacture, and it is extremely doubtful whether anything like present levels of food production could be maintained without its use for this purpose. World consumption of ammonia at the time of writing is about 65 million tonnes per annum. US capacity, mostly, but not all, based on steam reforming, is about 17.8 million tonnes per annum.

The object of steam reforming in an ammonia plant is to produce a mixture of nitrogen and hydrogen for use in the ammonia synthesis reaction:

$$N_2 + 3H_2 \longrightarrow 2NH_3$$

The nitrogen is obtained from the air, and the hydrogen arises partly from the hydrocarbon feed and partly from water. The carbon content of the hydrocarbon is rejected from the process as carbon dioxide. The steam reforming reaction as such forms only one step in a complex, multistage process (Fig. 9.1).

A number of the catalysts used in ammonia manufacture are poisoned by sulphur, so the first step in an ammonia process, as in all processes using steam reforming, is removal of sulphur-containing

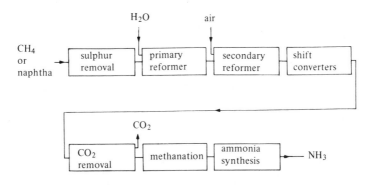

Fig. 9.1 – Steam reforming for ammonia manufacture.

compounds. In the case of natural gas the sulphur is present as hydrogen sulphide and thiols, and these are easily removed by passing the gas through beds of active carbon, or of zinc oxide at 350 to 400°C. Naphtha is hydrodesulphurised (see section 6.2).

The steam-reforming reaction is carried out in two stages, called primary and secondary reforming. In the **primary reformer**, the catalyst is held in tubes of about 10 cm diameter, heated in a furnace by burning gas or fuel oil. Steam and feed, in the ratio of about 3 kg steam to 1 kg of feed, and at a pressure of around 30 atm., are fed into the reformer, and reach a temperature in the region of 750 to 850°C. As in the case of thermal cracking, the maximum temperature that can be reached is determined by the maximum working temperature of the alloy of which the tubes are made. The residence time in the reformer is such that the reforming reactions do not go to completion, so that the exit gases still contain significant quantities of hydrocarbon, in addition to hydrogen, carbon monoxide, carbon dioxide, and steam.

The **secondary reformer** is a steel vessel lined with refractory brickwork, and containing a single bed of catalyst. In addition to the hot gases from the primary reformer, it is fed with air. This has two effects. Firstly, the resulting combustion of some of the gas raises the temperature to over 1000°C. At this temperature the basic reforming reaction goes virtually to completion, so that only a trace of methane remains. Secondly, it introduces the required nitrogen into the process stream. The gas leaving the secondary reformer contains nitrogen, hydrogen, carbon monoxide, carbon dioxide, steam, and a trace of methane. Apart from the ammonia synthesis stage itself, the rest of the process is concerned with the removal of carbon oxides from the gas. These are poisons for the ammonia synthesis catalyst, and have to be virtually entirely removed from the synthesis gas.

The question arises as to why reforming is carried out at an elevated pressure, since this will have an undesirable effect on the position of the equilibrium in the reforming reaction. The reason is that the synthesis gas eventually has to be compressed to a high pressure for the ammonia synthesis stage, and it is economically advantageous to carry out as much of the compression as possible on the smaller volume of feedstock.

From the reformers, the gas passes to the **shift converters**. The object here is to convert the carbon monoxide as completely as possible to carbon dioxide, via the shift reaction:

$$CO + H_2O \rightleftharpoons CO_2 + H_2$$

Since the shift reaction is exothermic, the position of the equilibrium will be more favourable to this purpose at low temperatures. On the other hand, the lower the temperature, the lower the rate of reaction will be. To achieve a balance between these conflicting requirements, shift conversion is normally carried out in two stages, the high temperature shift and the low temperature shift. Both are carried out in adiabatic reactors. The high temperature shift reactor contains an iron oxide catalyst, and operates at about 400°C. At this temperature, substantial amounts of carbon monoxide remain in the gas stream. The gas is then cooled to about 200°C, and passed to the low-temperature shift reactor which contains a copper–zinc catalyst. This is active at lower temperatures than the iron oxide catalyst, but on the other hand is more expensive. The exit stream from the low temperature shift reactor contains only a trace of carbon monoxide.

The next stage is **carbon dioxide removal**. This is most often carried out by scrubbing the gas with a solution of potassium carbonate or of ethanolamine:

$$CO_2 + H_2O + K_2CO_3 \rightleftharpoons 2KHCO_3$$

$$CO_2 + H_2O + H_2NCH_2CH_2OH \rightleftharpoons HCO_3^- \ H_3\overset{+}{N}CH_2CH_2OH$$

The solution emerging from the scrubber is heated to drive off carbon dioxide and regenerate the scrubbing solution. The gas stream leaving this stage contains nitrogen, hydrogen, and traces of carbon dioxide, carbon monoxide and methane, together with a small amount of water vapour.

Removal of the last traces of carbon oxides is carried out by the process of **methanation**. This involves passing the gas over a nickel catalyst at about 350°C. At this temperature, the methane reforming reaction lies far over towards starting materials, and consequently the carbon oxides are converted via this reaction, in conjunction with the shift reaction in the case of carbon dioxide, to methane and water:

$$CO + 3H_2 \longrightarrow CH_4 + H_2O$$

$$CO_2 + 4H_2 \longrightarrow CH_4 + 2H_2O$$

The gas stream from the methanator is cooled, and the water condenses and is separated to give a synthesis gas containing nitrogen and hydrogen in a molar ratio of $1:3$ together with a small amount of methane.

9.3 AMMONIA SYNTHESIS

The ammonia synthesis reaction is reversible:

$$N_2 + 3H_2 \rightleftharpoons 2NH_3$$

Since it is exothermic, the position of the equilibrium becomes less favourable the higher the temperature.

Synthesis is carried out over promoted iron catalysts. Typically, these require a temperature of at least 400°C for an adequate rate of reaction, and at this temperature the equilibrium constant is very small. To achieve a reasonable degree of conversion the reaction is carried out under pressure, commonly in the range 140 to 350 atm. The design of reactors, called 'converters', to operate under these conditions presents some problems. One particular feature is that the vessel walls cannot be at the reaction temperature, since at this temperature and under the pressures involved, 'hydrogen embrittlement' of the metal would occur. This problem is surmounted by arranging that the cold synthesis gas entering the converter flows through a space between the vessel wall and the catalyst, thus keeping the wall cool. The gas is then heated to synthesis temperature by an internal heat exchanger before contacting the catalyst. Various arrangements are used for providing cooling, for temperature control, in the converter. One method is to have the catalyst in a series of beds and to cool the gas stream between the beds. Temperatures vary with the particular plant design, but are commonly in the range 400 to 550°C. The conversion obtained varies, depending on the particular conditions of temperature and pressure used, but is often in the region 20% to 30%.

Ammonia is separated from the gas stream by cooling it to a temperature at which the ammonia liquefies. Usually, before this is done the incoming synthesis gas is added to the gas from the converter, so that in the condenser the make-up gas is, in effect, scrubbed with liquid ammonia. This removes any traces of carbon dioxide or water that remain in the synthesis gas.

The whole of the equipment involved in synthesis, comprising a compressor, converter, and condenser, is known as a *synthesis loop* (Fig. 9.2). Since the synthesis gas contains small amounts of methane, from the steam reforming, and argon, from the air, a small proportion of the gas in the loop has to be constantly withdrawn as a purge, to avoid the build-up of these inert materials in the system.

The development of a catalyst that would allow the synthesis reaction to be carried out at lower temperatures, and therefore lower pressures, would clearly reduce costs, and much research has been carried out in search of such a catalyst. In the main, this appears to have produced only marginal improvements. In 1982, ICI announced the development of a process which uses a synthesis pressure of only 70 to 80 atm. No details appear to have been released of the catalyst used or of the reaction temperature, and it remains to be seen how significant a development this is.

Modern ammonia plants operate on a similar scale to ethylene plants, capacities commonly being in the range 250 000 to 500 000 tonnes per annum.

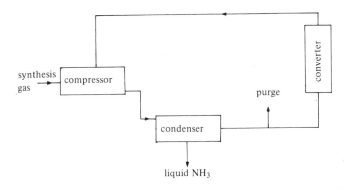

Fig. 9.2 — Ammonia synthesis loop (simplified).

9.4 USES OF AMMONIA

We have already seen a number of uses of ammonia, and of nitric acid which is derived from it, in making petrochemicals. However, by far its most important use, accounting for over 80% of consumption, is in fertiliser applications. Some ammonia is used as such, being applied to the soil either by direct injection as a gas or in aqueous solution, but most is converted into solid derivatives, notably ammonium nitrate, ammonium sulphate, and urea.

Nitric acid manufacture is based on oxidation of ammonia with air over a platinum or platinum–rhodium gauze catalyst at about 900°C. Nitric oxide is produced, and this then oxidises non-catalytically in air to nitrogen dioxide:

$$4NH_3 + O_2 \longrightarrow 4NO + 6H_2O$$

$$2NO + O_2 \longrightarrow 2NO_2$$

Absorption of nitrogen dioxide in water gives nitric acid:

$$3NO_2 + H_2O \longrightarrow 2HNO_3 + NO$$

It is then necessary for the nitric oxide formed at this stage to be oxidised back to nitrogen dioxide by air. Consequently, the absorption has effectively to be carried out in a large number of stages with air oxidation of the gas stream in between each stage. This is achieved by using large towers filled with trays or packing to promote gas–liquid contact. The yield of nitric acid is about 95%.

Urea is made by reacting carbon dioxide and ammonia at 175 to 220°C and 170 to 400 atm.:

$$2NH_3 + CO_2 \longrightarrow NH_2CONH_2 + H_2O$$

The carbon dioxide produced by the ammonia plant is normally used for this process.

Urea is used as a fertiliser, as an animal feed supplement, and as a chemical intermediate, notably in the manufacture of urea–formaldehyde resins and of melamine.

9.5 METHANOL – MANUFACTURE

Methanol plants are much less complicated than ammonia plants. They involve only desulphurisation, reforming, and methanol synthesis. The reforming reactions are carried out entirely in a primary reformer. Consequently, the maximum reforming temperature is substantially less than in an ammonia plant, and the methane content of the reformed gas is somewhat higher.

The methanol synthesis reaction, like the ammonia synthesis reaction, is reversible and exothermic:

$$CO + 2H_2 \rightleftharpoons CH_3OH$$

For many years, the catalysts used for this reaction were based on promoted zinc oxide. Catalysts of this type require a reaction temperature of 300 to 400°C, and at these temperatures a pressure of around 300 atm. has to be used to achieve an acceptable conversion. It can be seen that these conditions are similar to those in ammonia synthesis, and in fact the reactors used are virtually the same as ammonia converters.

In the early 1960s, ICI developed a process which uses a copper-based catalyst for the synthesis reaction. This is much more active than the zinc oxide catalysts, and will operate at reaction temperatures of 200 to 300°C. At these temperatures, acceptable conversions can be obtained at pressures of 50 to 100 atm., and this allows major savings in operating and capital costs over the high-pressure process. Notable features are that the converter walls can be at reaction temperature, so that its construction is much simpler, and that a much smaller compressor is required. Since 1966, when the first ICI low-pressure plant went into operation, the vast majority of new methanol plants that have been built have used the low pressure process.

The synthesis reaction requires a molar ratio of hydrogen to carbon monoxide of 2:1, whereas steam reforming of methane gives a gas with a ratio of 3:1. Commonly the ratio is adjusted by adding carbon dioxide to the reformer feed. This is then converted in the reformer to carbon monoxide, via the reverse shift reaction:

$$CO_2 + H_2 \longrightarrow CO + H_2O$$

If the methanol plant is situated near an ammonia plant, the carbon dioxide effluent from this can be used. Otherwise, carbon dioxide may be recovered from the reformer furnace flue gases.

Typically in the low-pressure process, the converter contains a single bed of catalyst, and cooling is achieved by injecting cold synthesis gas into the bed at a number of points along its length. The converter is incorporated into a synthesis loop of similar overall configuration to an ammonia synthesis loop. The reacted gases are cooled to condense out the methanol, and the remaining gases are recirculated to the converter, a purge being taken off to avoid the build-up of inert materials, that is, methane and sometimes nitrogen from the natural gas. The purge stream is burnt as fuel in the reformer furnace. Methanol is purified by distillation.

US capacity for methanol is about 4.7 million tonnes per annum.

9.6 METHANOL – USES AND DERIVATIVES

Since the oil crises in 1973 there has been a great deal of research and development interest in routes to organic chemicals based on raw materials other than petroleum fractions. Much of this interest has been centred around methanol, which can be made from synthesis gas from methane, as we have seen, and also from coal (section 9.8.1), and a number of new processes have been developed. At the time of writing, most of these processes have not progressed beyond the pilot-plant stage, and old-established outlets for methanol account for most of its consumption.

9.6.1 Formaldehyde

The manufacture of formaldehyde has traditionally been the major outlet for methanol. It accounts for about 40% of consumption.

Methanol may be converted into formaldehyde either by dehydrogenation, which is endothermic, or by oxidation, which is exothermic:

$$CH_3OH \longrightarrow HCHO + H_2$$

$$CH_3OH + O_2 \longrightarrow HCHO + H_2O$$

Most manufacture is carried out by a combined oxidation–dehydrogenation process in which the heat of reaction for the dehydrogenation is provided by the oxidation. Typically, an air–methanol mixture, with the methanol in substantial excess so that the reaction mixture stays outside the higher explosive limit, is passed through a shallow bed of silver crystals or a number of layers of silver gauze. The reaction temperature is about 600 to 650°C. The gases emerging from the reactor are rapidly cooled, and are then passed to an absorber where the formaldehyde and residual methanol are dissolved in water. The resulting solution is distilled to recover methanol, which is recycled, and an aqueous solution of formaldehyde. The yield is in the range 86% to 90%. Formaldehyde is stored and transported in aqueous solution, most commonly of 37% concentration.

US capacity for formaldehyde is about 1.5 million tonnes per annum.

Formaldehyde has a wide range of uses, some of which we have met. The most important in terms of tonnage are in the manufacture of phenol–formaldehyde resins and amino resins.

9.6.2 Acetic acid

We have seen that acetic acid is made by the oxidation of acetaldehyde (section 3.6.1). It is also made by oxidation of butane and of naphtha (section 10.3). However, since about 1970, most new capacity has been based on the carbonylation of methanol:

$$CH_3OH + CO \longrightarrow CH_3CO_2H$$

There are two commercial processes based on this reaction. The first to be operated, developed by the German company BASF, uses a catalyst system containing cobalt, copper, and iodine. Reaction is carried out in the liquid phase at about 250°C and under about 650 atm. pressure. The yield of acetic acid is about 90% based on methanol and about 60% based on carbon monoxide. This process has been operated to only a limited extent.

The process which is now the favoured one for new acetic acid plants was developed by Monsanto and first put into operation in 1970. This uses a catalyst system containing rhodium and iodine, and operates at 150 to 200°C and 30 to 40 atm., on a feed of methanol containing 30% to 40% water. Methanol is completely converted, and the yield of acetic acid is about 99% based on methanol and 90% based on carbon monoxide. It is clear that both in terms of yield and operating conditions this process is much more attractive than the cobalt catalysed carbonylation.

It is thought that the active catalytic species in the Monsanto process is $Rh(CO)_2I_2$, and that the acetic acid is formed by the following sequence of steps:

$$[Rh(CO)_2I_2]^- + CH_3I \longrightarrow [CH_3Rh(CO)_2I_3]^-$$

$$\longrightarrow [CH_3CORhCOI_3]^- \xrightarrow{CO} [CH_3CORh(CO)_2I_3]^-$$

$$\xrightarrow{H_2O} CH_3CO_2H + [Rh(CO)_2I_2]^- + HI$$

$$CH_3OH + HI \longrightarrow CH_3I + H_2O$$

Manufacture of acetic acid consumes about 10% of methanol. US capacity for acetic acid (all processes) is about 1.85 million tonnes per annum.

The main use of acetic acid is in making acetates, either direct, or via acetic anhydride. Manufacture of vinyl acetate (section 3.7) is the largest such outlet, accounting for around 40% of consumption. Production of cellulose acetate, for use as a fibre and plastic, is the next most important outlet. For this application, the acetic acid is converted into acetic anhydride:

$$CH_3CO_2H \xrightarrow[700-750°C]{\text{triethyl phosphate}} CH_2{=}C{=}O + H_2O$$

$$\textit{ketene}$$

$$CH_2{=}C{=}O + CH_3CO_2H \xrightarrow{45-55°C} \begin{array}{c} CH_3C\diagup{}^O \\ \diagdown O \\ CH_3C\diagdown_O \end{array}$$

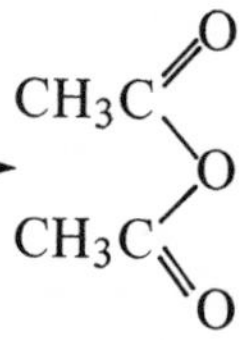

Various alkyl acetates are also made in significant quantities.

Quite substantial amounts of acetic acid are used as a solvent, notably in the oxidation of p-xylene to terephthalic acid.

9.6.3 Methylamines

Methylamines are made by reacting methanol with ammonia at temperatures in the range 350 to 500°C and about 20 atm., over a catalyst such as alumina:

$$CH_3OH \xrightarrow{NH_3} CH_3NH_2 \xrightarrow{CH_3OH} (CH_3)_2NH \xrightarrow{CH_3OH} (CH_3)_3N$$

The proportion of the products can be adjusted to some extent by varying the ratio of ammonia to methanol, but it is not feasible to avoid the formation of significant amounts of trimethylamine, for which there is only a small demand. However, net production of trimethylamine by the plant can be avoided by recycling it to be reactor.

About 4% of methanol is consumed in this application. The methylamines have a range of small to medium tonnage uses as chemical intermediates.

9.6.4 Methyl chloride
Rather more than 3% of methanol is used for the manufacture of methyl chloride:

$$CH_3OH + HCl \longrightarrow CH_3Cl + H_2O$$

Both liquid phase and gas phase operation are used.

In liquid phase processes, methanol and hydrogen chloride are passed into a concentrated aqueous solution of zinc chloride or iron(III) chloride at about 140°C. The methyl chloride, together with some unconverted methanol and hydrogen chloride pass out of the reactor in the gas phase. Methanol is recovered by scrubbing with water, and the gas stream is then scrubbed with concentrated sulphuric acid, which both dries it and removes by-product dimethyl ether. Methyl chloride is condensed by compressing the gas stream to about 8 atm.

In gas phase operation, reaction is carried out at 300 to 350°C. A variety of materials are used as catalyst, e.g. copper(II) chloride or zinc chloride on solid supports. Yields in both liquid and gas phase processes are around 95%.

Methyl chloride is also made by chlorinating methane. This process, and uses of methyl chloride, are discussed in section 10.1.

9.6.5 Other uses of methanol
We have discussed three other important uses of methanol as an intermediate in previous chapters, namely the manufacture of methyl methacrylate (section 4.2), methyl *t*-butyl ether (section 5.4), and dimethyl terephthalate (section 8.5). These consume about 3%, 10% and 4% of methanol respectively, at the time of writing. It is likley that the manufacture of methyl *t*-butyl ether will increase in relative

importance.

It has already been indicated that in recent years a number of new processes based on methanol have been developed. Some of these are outlined in Fig. 9.3. At present, the only such process which appears to have been put into commercial operation is that for the manufacture of acetic anhydride. Eastman Kodak are reported to have built a plant in the USA to make 450 000 tonnes per annum of acetic anhydride from methanol produced from synthesis gas made from coal.

In addition to its use as a chemical intermediate, a considerable amount of attention has recently been paid to the potential uses of methanol in fuel applications, either direct, or after further processing. Blending into gasoline is one possibility, and has been used to some extent. However, there are problems in that such blends can take up water from the air and this can result in the separating out of a methanol phase. Also, there have been suggestions that methanol–gasoline blends give rise to various corrosion problems. We have already seen that methyl *t*-butyl ether is being used to an increasing extent in gasoline, so this represents an indirect fuel use of methanol.

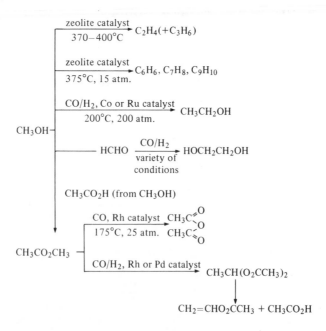

Fig. 9.3 — New processes based on methanol.

Another approach, being applied in New Zealand, is to convert methanol to gasoline over a zeolite catalyst. This process, which is essentially the same as that shown in Fig. 9.3 for the production of aromatic hydrocarbons, was developed by Mobil, and gives a gasoline of high aromatics content. It is attractive in New Zealand because they have abundant supplies of natural gas but no significant resources of crude oil.

The extent to which fuel uses of methanol develop will be determined by relative prices of petroleum, natural gas, and coal. In the long term, as oil reserves become depleted, it will presumably become necessary to manufacture liquid fuels from coal, and the synthesis gas/methanol route is one way of doing this. In the shorter term, in specific locations low price natural gas or coal may make the manufacture of methanol for fuel use economically attractive. We have already seen that New Zealand appears to be one region where this is the case.

One further use of methanol which at present consumes only small quantities, but which might become of major importance at some time in the future, is in the manufacture of single cell protein by fermentation. This process is currently operated by ICI at a scale of 60 000 tonnes per annum. The product is used as an animal feed.

9.7 STEAM REFORMING FOR OTHER PURPOSES

For some processes, notably the Oxo process, approximately equimolar mixtures of carbon monoxide and hydrogen are required. These are conveniently made by steam reforming a mixture of methane, or naphtha, and carbon dioxide, in the appropriate proportions.

Hydrogen is required for a wide variety of chemical applications. A number of processes, e.g. catalytic reforming and alkali manufacture, produce hydrogen in substantial quantities as a co-product, and hydrogen-using process are often situated where such a supply is available. When co-product hydrogen is not available, steam reforming is the usual method of generation.

A hydrogen plant comprises a primary reformer, shift converters, carbon dioxide removal, a methanator, and, where required, a hydrogen purification stage. Typically, a gas stream containing about 98% hydrogen is obtained from the methanator, the rest being largely methane, and this is suitable for many purposes. When a purer product is required, the gas stream is subjected to cryogenic separation or, more usually these days, to *pressure swing adsorption*. This involves passing the gas through a bed of solid adsorbent which takes out the methane and any other impurities. The bed is periodically regenerated by purging with hydrogen at reduced pressure.

Where **carbon monoxide** alone is required it is separated from the reformer product either by cryogenic separation or by selective absorption. When using the former method, the reformed gas must be freed of carbon dioxide, e.g. by scrubbing with ethanolamine solution, and then dried by passing through a bed of molecular sieve. The gas is then cooled to condense out carbon monoxide and methane, and the carbon monoxide is separated by fractional distillation. Separation by absorption involves scrubbing the gas with either an aqueous solution of a copper(I) salt, or a solution of copper(I) chloride and aluminium chloride in toluene, when a carbon monoxide–copper(I) complex is formed. Increasing the temperature and reducing the pressure decomposes the complex, and releases the carbon monoxide.

9.8 SYNTHESIS GAS BY OTHER METHODS

9.8.1 From coal

Manufacture of synthesis gas was based on coal for many years before the introduction of steam reforming, and in Western Europe coal-based production remained dominant until the 1960s. The original method was generation from coke, by the water gas reaction:

$$C + H_2O \xrightarrow[c.\ 1200°C]{} CO + H_2$$

The heat required for this endothermic reaction is provided by burning some of the coke:

$$C + O_2 \longrightarrow CO_2$$

Operation is cyclic, with air and steam alternately being blown through a bed of coke. The process is extremely wasteful of coke and of energy. More sophisticated and efficient are coal gasification processes, in which coal is treated with mixtures of oxygen, or air, and steam:

$$\text{coal} \xrightarrow[750-1500°C]{O_2, H_2O} CO + H_2$$

Capital and operating costs for coal-based manufacture of synthesis gas are much higher than those for manufacture by steam reforming, largely because coal is a solid. During the era of cheap oil and natural gas which extended up to 1973, manufacture of synthesis gas by steam reforming was much more economically attractive than manufacture from coal. Since 1973, the price of oil and natural gas has increased to such an extent that in some areas where low cost coal is available, e.g. in some parts of the USA, it may now be attractive to move back to the use of coal as a raw material for the production of synthesis gas. It remains to be seen to what extent this happens in the short to medium term. In the long term, it is highly likely that it will happen, since reserves of coal are much greater than reserves of oil and natural gas.

There is one country, South Africa, where manufacture of synthesis gas from coal has never been displaced by steam reforming. There are very specific reasons for this. South Africa has no substantial reserves of oil or natural gas, but on the other hand has large reserves of easily accessible coal and plenty of cheap labour. It also has a strong political desire to remain as independent as possible of imported oil.

9.8.2 From petroleum fractions by partial oxidation

Synthesis gas can be made from hydrocarbon feeds by reaction with a limited supply of oxygen or oxygen enriched air. Reaction temperatures are typically in the range 1100 to 1500°C, and pressures are around 20 atm.; no catalyst is used. Using methane as an example (but see below) the major gas-forming reaction may be represented as:

$$CH_4 + \tfrac{1}{2}O_2 \longrightarrow CO + 2H_2$$

Partial oxidation is seldom, if ever, applied using natural gas or naphtha as feedstock, since, for a variety of reasons, including the fact that it requires an air-separation plant to provide oxygen, the capital and operating costs involved are higher than for steam reforming. It is used, to a limited extent, to manufacture synthesis gas from atmospheric residue. Atmospheric residue (section 1.4) is cheap, but is unsuitable for use as a feedstock for steam reforming.

PROBLEMS AND EXERCISES

1. Draw an outline flow scheme for the carbon dioxide removal section of an ammonia plant.

2. One of the disadvantages of using naphtha as a feedstock for ammonia manufacture is that carbon dioxide removal costs are higher than when methane is used. Why is this so? (Amongst other disadvantages are that the naphtha has to be vaporised, that sulphur removal is more difficult, and that more complicated reforming catalysts are required.)

3. Draw up a list of petrochemical processes which use (i) ammonia, and (ii) nitric acid as raw materials.

4. US consumption of methanol in 1983 was about 3.35 million tonnes. Estimate what proportion of acetic acid was made from methanol in the USA that year, given that total acetic acid production was about 1.3 million tonnes.

5. There is much less concentration of manufacture of formaldehyde into a small number of large plants than there is for most petro-chemicals; many users of large amounts of formaldehyde make their own from purchased methanol. Can you suggest why formaldehyde is atypical in this way?

6. Draw an outline flow scheme for hydrogen manufacture by steam reforming.

10

Miscellaneous processes
and products

We have seen that the majority of petrochemical products are made from basic intermediates produced by the three building block processes, thermal cracking, catalytic reforming, and steam reforming. However, there are a number of processes and products that do not fit into this pattern. This chapter deals with the most important of these.

10.1 CHLOROMETHANES

The chloromethanes are produced commercially on a fairly substantial scale (Table 10.1). Chlorination of methane is one method of manufacture, but all are also made by other routes.

Table 10.1

US capacities for chloromethanes

	Capacity (tonnes per annum)
Methyl chloride (CH_3Cl)	312 000
Methylene chloride (CH_2Cl_2)	324 000
Chloroform ($CHCl_3$)	178 000
Carbon tetrachloride (CCl_4)	367 000

Chlorination of methane is usually carried out at 400 to 450°C. It is a free radical chain reaction. Initiation is by thermal scission of chlorine molecules, and the propagation reactions (for methane) are as follows:

$$Cl\cdot + CH_4 \longrightarrow HCl + CH_3\cdot$$

$$CH_3\cdot + Cl_2 \longrightarrow CH_3Cl + Cl\cdot$$

Since methyl chloride, methylene chloride, and chloroform are all also attacked under the reaction conditions, a mixture of products is formed:

$$CH_4 \xrightarrow{Cl_2} \underset{+\ HCl}{CH_3Cl} \xrightarrow{Cl_2} \underset{+\ HCl}{CH_2Cl_2} \xrightarrow{Cl_2} \underset{+\ HCl}{CHCl_3} \xrightarrow{Cl_2} \underset{+\ HCl}{CCl_4}$$

Small amounts of products containing more than one carbon atom arise from termination reactions.

The reaction is very exothermic, and proceeds so rapidly that conventional methods of heat transfer are inadequate for temperature control. Commonly, a substantial excess of methane is used, or carbon tetrachloride is recycled, so that the thermal capacity of the reacting mixture controls the temperature in the desired range. The product gases are cooled, washed with water to remove hydrogen chloride, scrubbed with sodium hydroxide solution, and cooled to liquefy the chloromethanes. Unconverted methane is recycled, and the chloromethanes are separated by distillation.

The product distribution can be controlled to some extent by adjusting the ratio of methane to chlorine. If desired, methyl chloride, methylene chloride, and chloroform can be recycled to the reaction stage to increase the yield of the more highly chlorinated products.

A major factor in the economics of the process is the value of the co-product hydrogen chloride. The process would not normally be attractive where there is no outlet for the hydrogen chloride.

We have already seen (section 9.6.4) that methyl chloride is also made from methanol, and in fact this route accounts for the bulk of manufacture. Methylene chloride, together with a certain amount of chloroform and carbon tetrachloride, may be made by chlorination of methyl chloride. In this way, a methylene chloride plant based on methanol and chlorine can be operated with no major production of hydrogen chloride:

$$CH_3OH + HCl \longrightarrow CH_3Cl \xrightarrow{Cl_2} CH_2Cl_2 + HCl$$

Carbon tetrachloride is also made by chlorination of carbon disulphide (see next section) and, together with perchloroethylene, by **chlorinolysis** of various hydrocarbons and chlorohydrocarbons. Chlorinolysis involves reaction of the feed, e.g. propylene, with chlorine at about 600°C:

$$C_3H_6 \xrightarrow{Cl_2} CCl_4, CCl_2{=}CCl_2, HCl$$

Carbon tetrachloride and perchloroethylene are rapidly interconverted under the process conditions:

$$2CCl_4 \rightleftharpoons CCl_2{=}CCl_2 + 2Cl_2$$

Consequently, the output of the process can be adjusted to suit market requirements by recycling unwanted product.

The main use of methyl chloride is as a methylating agent, notably in making tetramethyl-lead for use as a gasoline antiknock additive (cf. tetraethyl-lead, section 3.9), and methylsilanes for silicone manufacture:

$$CH_3Cl + Si \xrightarrow[250-600°C]{Cu\ catalyst} (CH_3)_2SiCl_2,\ CH_3SiCl_3,\ (CH_3)_3SiCl$$

Practically all methylene chloride is used as a solvent. Important applications are in paint strippers, solvent extraction, and in spinning cellulose triacetate fibres.

The main use of chloroform is in making chlorodifluoromethane, an important refrigerant fluid:

$$CHCl_3 + 2HF \longrightarrow CHClF_2 + 2HCl$$

It has some other applications as a chemical intermediate, and is used to some extent as a solvent. Its original use, as an anaesthetic, has to all intents and purposes been discontinued because of its toxicity.

Carbon tetrachloride is mainly used in making dichlorodifluoromethane, and trichlorofluoromethane, for use as refrigerants and aerosol propellants:

$$CCl_4 \xrightarrow{HF} CCl_3F,\ CCl_2F_2,\ HCl$$

It is also used to some extent as a solvent, and in fire extinguishers.

10.2 CARBON DISULPHIDE

Carbon disulphide was formerly made by the reaction of charcoal with sulphur. This process has now largely been replaced by manufacture based on the reaction of methane and sulphur:

$$CH_4 + 4S \longrightarrow CS_2 + 2H_2S$$

The reaction is carried out at about 600°C and 2 to 5 atm., over a catalyst (often alumina or clay). Sulphur is used in slight excess, and the conversion of methane is around 95%. The exit gases are cooled, and scrubbed with oil to remove the carbon disulphide which is subsequently recovered by fractional distillation. Excess sulphur is recycled, and the hydrogen sulphide is treated in a Claus kiln for sulphur recovery:

$$2H_2S + 3O_2 \longrightarrow 2SO_2 + 2H_2O$$

$$2H_2S + SO_2 \longrightarrow 3S + 2H_2O$$

The yield of carbon disulphide, based on methane, is 85% to 90%.

The main use of carbon disulphide, accounting for more than 50% of production, is in making viscose rayon and regenerated cellulose film. Substantial amounts are also used in making carbon tetrachloride:

$$CS_2 + 3Cl_2 \xrightarrow[30°C]{Fe} CCl_4 + S_2Cl_2$$

$$CS_2 + 2S_2Cl_2 \longrightarrow CCl_4 + 6S$$

In addition to the above, there are a variety of other uses of carbon disulphide as a chemical intermediate.

10.3 ACETIC ACID BY BUTANE AND NAPHTHA OXIDATION

We have seen that acetic acid is made by oxidation of acetaldehyde (section 3.6.1), and by carbonylation of methanol (section 9.6.2), the latter being the favoured process for new plants. Substantial quantities are also made by liquid phase free radical oxidation of *n*-butane and of naphtha.

Oxidation of *n*-butane was first operated by Celanese in the USA in 1952. Typically it is carried out at 170 to 180°C and under about 50 atm. pressure, with cobalt acetate in solution as a catalyst. A complex mixture of products is obtained, and separated by fractional distillation. The yield of acetic acid is about 50%.

From what we have seen in previous chapters, it can be appreciated that a very large number of reactions can occur in this system. Reaction paths resulting in the formation of acetic acid can readily be proposed:

$$CH_3CH_2CH_2CH_3 \longrightarrow CH_3\overset{\overset{\displaystyle OOH}{|}}{C}HCH_2CH_3 \longrightarrow CH_3\overset{\overset{\displaystyle O^\cdot}{|}}{C}HCH_2CH_3$$

$$CH_3\overset{\overset{\displaystyle O}{\|}}{C}H \xrightarrow{\text{ further oxdn. }} CH_3CO_2H$$

$$\longrightarrow \quad +$$

$$CH_3CH_2^\cdot \longrightarrow CH_3CH_2OOH$$

$$\longrightarrow CH_3CH_2O^\cdot \longrightarrow CH_3CHO$$

However, many other reaction paths, resulting in other products, appear to be just as plausible, so that at first sight it seems rather surprising that a fairly high yield of acetic acid can be obtained. There is a very specific reason for this: acetic acid is quite stable to free radical oxidation, whereas the other initial products are not. Consequently, the process conditions can be arranged so that other products tend to undergo further reaction, and with many of them this gives rise to further production of acetic acid.

Light naphtha can also be oxidised to acetic acid under similar conditions to those used for butane oxidation, though not surprisingly the yield is lower than than from butane. This process was developed by the Distillers Company Ltd. in the UK, and is operated in this country by BP at a scale of 200 000 tonnes per annum.

Although by-product production from these processes can be minimised by appropriate design of reaction conditions, and by recycling unwanted materials, it cannot be totally eliminated, and acetic acid plants using this type of process give significant amounts of by-products. Acetone is the most important of these; a light naphtha oxidation plant produces about 0.33 tonnes of acetone for every tonne of acetic acid.

10.4 DEHYDROGENATION AND CHLORINATION OF LINEAR ALKANES

We saw in section 7.6 that linear sodium alkylbenzenesulphonates are of major importance in domestic detergents. Linear 1-alkenes produced by oligomerisation of ethylene (section 3.10) are extensively used as alkylating agents in their manufacture, but chloroalkanes and alkenes produced from linear alkanes are also of substantial importance, as are alkenes produced by wax cracking (section 10.5).

Linear alkanes from C_{10} to C_{14} are separated from a kerosine fraction of appropriate boiling range, either by adsorption on a zeolite with a pore size which will admit straight chain alkanes but not branched ones, or by complex formation with urea. (Urea forms crystalline complexes with straight chain alkanes, but not with branched ones.) Chlorination of the alkanes gives a mixture of monochloroakanes, e.g.:

$$C_{12}H_{26} \xrightarrow[c.\ 100°C]{Cl_2} C_{12}H_{25}Cl + HCl$$

Conversion is limited to about 20% to avoid polychlorination. The chloroalkanes may be used as such to produce detergent alkylate, or alternatively may be catalytically dehydrochlorinated to give a mixture of linear alkenes. More commonly, linear alkenes are produced from the alkanes by catalytic dehydrogenation, e.g.:

$$C_{12}H_{26} \xrightarrow[300-500°C]{catalyst} C_{12}H_{24} + H_2$$

Conversion is limited to 10% to 15%, to avoid the formation of dienes. The resulting mixture of alkenes and alkanes may be used without separation in the alkylation step, and the alkanes, after separation from the detergent alkylate, recycled to the dehydrogenation.

10.5 WAX CRACKING

Thermal cracking of paraffin wax is used to produce linear 1-alkenes from C_6 to about C_{20}. We have already seen that 1-alkenes in this range are also made by oligomerisation of ethylene and have a number of uses, notably in the manufacture of plasticiser alcohols and of surface active agents. Wax cracking, the older-established process, is tending to diminish in relative importance.

Paraffin wax is a mixture of mainly *n*-alkanes from about C_{18} to C_{56}. It crystallises out from the gas oil and vacuum distillate fractions from oil refining, and is separated by filtration. Typically, it is cracked at 540 to 565°C and 2 to 4 atm. pressure in the presence of a small amount of steam for 5 to 15 seconds, to produce a mixture of straight chain 1-alkenes from ethylene upwards. The higher alkenes are recycled, and the rest are separated into a number of fractions appropriate to the various applications. The yield of C_6 to C_{20} alkenes is about 60%.

It is interesting to compare this process with naphtha cracking, which gives ethylene as the major product. Both processes involve the same types of reaction, but the lower temperatures and higher partial pressures used in wax cracking result in chain transfer and termination reactions being much faster relative to β-scisson reactions than in naphtha cracking. A typical sequence of reactions for a molecule of alkane is:

$$C_nH_{2n+2} \xrightarrow{\ -[H\cdot]\ } \sim\!\!\sim\!\!\sim\!\!\dot{C}HCH_2CH_2\!\!\sim\!\!\sim\!\!\sim$$

$$\xrightarrow{\ \beta\text{-scission}\ } \sim\!\!\sim\!\!CH{=}CH_2 + \cdot CH_2\!\!\sim\!\!\sim\!\!\sim$$

$$\sim\!\!\sim\!\!CH_2\cdot \left\{ \begin{array}{l} \xrightarrow{\ C_nH_{2n+2}\ } \sim\!\!\sim\!\!CH_3 + \sim\!\!\sim\!\!\dot{C}HCH_2CH_2\!\!\sim\!\!\sim \\ \xrightarrow{\ \cdot CH_2\!\!\sim\!\!\sim\ } \sim\!\!\sim\!\!CH_2CH_2\!\!\sim\!\!\sim \end{array} \right.$$

Under naphtha-cracking conditions, the primary radical formed in the first β-scission would usually undergo a sequence of β-scissions, each giving rise to a molecule of ethylene.

10.6 CRACKING FOR ACETYLENE

The first use of acetylene for organic chemical manufacture was in 1908, for the production of trichloroethylene. Manufacture of acetaldehyde and derived products began in 1916, and by the end of the 1930s acetylene was established as a key intermediate in the organic chemical industry, particularly in Germany.

These developments were based on the manufacture of acetylene from coke and lime, via calcium carbide, a cumbersome route involving a lot of solid handling, very high energy consumption, and the production of large amounts of waste calcium hydroxide. The evident shortcomings of the carbide route stimulated efforts to find other routes to acetylene, and during World War II some acetylene was produced in Germany by hydrocarbon cracking. After the war, as the petrochemical industry began to spread to all developed countries, intense efforts to develop economical petrochemical routes to acetylene were made, and a number of plants were built. However, for reasons discussed below, the acetylene produced by these processes was still much more expensive than other petro- chemical intermediates such as ethylene and propylene, and during the 1960s and early 1970s, acetylene-based routes were inexorably displaced by other routes. Acetylene is now, in general, of only very minor importance as an intermediate for organic chemical manu- facture. The one exception to this is South Africa, where acetylene, made from carbide, is still used on a large scale. We have already discussed reasons why South Africa wishes to avoid the use of oil as a raw material so far as possible.

Acetylene can be made from hydrocarbons, from methane upwards, by cracking at very high temperatures (1200°C or more) for very short times (less than 0.1 sec). These conditions present major engineering problems, requiring as they do very rapid heat transfer into and out of the process streams at very high temperatures. A variety of approaches have been used, for example, heating the feed in an electric arc, or in a flame, or by contact with hot brick- work. In all cases the capital cost is high compared with the cost of an ethylene cracking furnace. The energy costs are also high in consequence of the high temperatures involved, and the fact that the necessity for a rapid quenching of the cracked gases means that very little energy can be recovered. In addition to the problems associated with the cracking stage, the costs involved in separating acetylene from the cracked gases are much higher than the separation costs in an ethylene plant. Liquid acetylene is highly explosive, and consequently separation has to be by solvent extraction rather than by distillation.

PROBLEMS AND EXERCISES

1. Draw an outline flow scheme for a plant to make carbon tetra-chloride alone by chlorinolysis of propylene.
2. Methyl ethyl ketone and propionic acid (propanoic acid) are two of the by-products formed in the liquid phase air oxidation of butane. Suggest mechanisms for their formation.

3. Draw up a list of petrochemicals that were formerly made from acetylene. What is the most important process currently in use for each of these?

4. Why can acetylene not be made in a tubular furnace similar to an ethylene cracking furnace?

Sources of further information and further reading

There is a lot of information available about petrochemicals, but it is very widely dispersed. The following list is not by any means intended to be comprehensive.

GENERAL REFERENCE WORKS

Kirk-Othmer Encyclopaedia of Chemical Technology, 3rd edn., John Wiley, New York, Vol. 1 (1978) to Vol. 24 (1984).
[An invaluable source of information.]

Encyclopaedia of Chemical Processing and Design, J. K. McKetta (Ed.), Marcel Dekker Inc., New York, Vol. 1 (1976) to Vol. 19 (Eth.) (1983) available at the time of writing.
[Another invaluable work, though at the present rate of progress it will be many years before it is complete.]

Ullmann's Encyclopaedia of Industrial Chemistry, Verlag Chemie. 36 volumes, to be published at the rate of 3 to 4 volumes a year, starting from 1984.
[Similar in scope to Kirk-Othmer, though with rather more emphasis on chemistry.]

Faith, Keyes and Clark's Industrial Chemicals, F. A. Lowenheim and M. K. Moran, 4th edn., John Wiley, New York (1975).

[A compact source of a great deal of information about processes and products. The fourth edition is now becoming rather out of date. A fifth edition, if it appears, should be extremely useful.]

GENERAL TEXTBOOKS

Basic Organic Chemistry, Part 5, *Industrial Products*, J. M. Tedder, A. Nechvatal and A. H. Jubb, John Wiley, London (1975).

Industrial Organic Chemistry, K. Weissermel and H.-J. Arpe, Verlag Chemie, New York (1978).

Chemicals from Petroleum, A. L. Waddams, 4th edn., Murray, London (1978).

An Introduction to Industrial Organic Chemistry, P. Wiseman, 2nd edn., Applied Science, London (1979).

Organic Chemicals in Perspective, Vols. 1 and 2, H. A. Wittcoff and B. G. Reuben, John Wiley, New York (1980).

Catalysis and Chemical Processes, R. Pearce and W. R. Patterson, Leonard Hill, Glasgow (1981).

From Hydrocarbons to Petrochemicals, L. F. Hatch and M. Matar, Gulf Publishing, Houston (1981).

BOOKS DEALING WITH SPECIFIC AREAS

Modern Petroleum Technology, G. D. Hobson, 5th edn., John Wiley, Chichester (1984).

Our Industry Petroleum, The British Petroleum Company Ltd., 5th edn., London (1977). Chapter 12 – crude oil; Chapter 13 – refining; Chapter 18 – natural gas.

The Chemical Economy, B. G. Reuben and M. L. Burstall, Longman, London (1973). Chapter 15 – process technology.

Ethylene and its Industrial Derivatives, S. A. Miller, Benn, London (1969).

Ethylene: Basic Chemicals Feedstock Material, O. G. Farah, R. P. Oullette, R. C. Kuehnel, M. A. Muradaz and P. N. Cheremisinoff, Ann Arbor Science Publishers Inc., Ann Arbor Michigan (1980).

Ethylene: Keystone to the Petrochemical Industry, L. Knied, O. Winter and K. Stork, Marcel Dekker, New York (1980).

Pyrolysis: Theory and Industrial Practice, L. F. Albright, B. L. Crynes and W. H. Corcoran, Academic Press, New York (1983).

Chemistry of Catalytic Processes, B. C. Gates, J. R. Katzer and G. C. A. Schuit, McGraw–Hill, New York (1979). Chapter 2 – ethylene oxide, vinyl acetate, Ziegler–Natta polymerisation, Oxo process; Chapter 3 – catalytic reforming; Chapter 4 – acrylonitrile.

Propylene and its Industrial Derivatives, E. G. Hancock, Benn, London (1973).

Benzene and its Industrial Derivatives, E. G. Hancock, Benn, London (1975).

Toluene, the Xylenes and their Industrial Derivatives, E. G. Hancock, Elsevier, Amsterdam (1982).

Catalyst Handbook, Imperial Chemical Industries Ltd., Wolfe Scientific Books, London (1970).

[Steam reforming and ammonia synthesis.]

Chemicals from Synthesis Gas, R. A. Sheldon, B. Reidel Publishing Company, Dordrecht (1983).

[Includes a good discussion of the Oxo process.]

JOURNALS

The following is a selection of journals which regularly contain useful material about petrochemicals.

Chemical and Engineering News
Chemical and Engineering Progress
Chemical Technology
Chemistry in Britain
Chemistry and Industry
European Chemical News
Hydrocarbon Processing
Oil and Gas Journal

Answers to problems

CHAPTER 1

1. All C_6 alkanes, methylcyclopentane, cyclohexane, benzene.
3. 36%. No.
4. (i) 60%, (ii) 84%, (iii) 82%.

CHAPTER 2

1. H_2, 1.1; CH_4, 0.15; C_2H_4, 1.0. Proportion of hydrogen would decrease and of methane would increase.
2. 86%.
3. $PhCH_2CH_2CH_2CH_3 \longrightarrow Ph\overset{\cdot}{C}HCH_2CH_2CH_3$
   $\longrightarrow PhCH=CH_2 + CH_3CH_2\cdot$

   $PhCH_2CH_2CH_2CH_3 \longrightarrow PhCH_2CH_2\overset{\cdot}{C}HCH_3$
   $\longrightarrow CH_2=CHCH_3 + PhCH_2\cdot \longrightarrow PhCH_3$

4. (i) Propylene, some isobutene, methane, some hydrogen; (ii) isobutene and methane.
5. Ethylene formation occurs only from ions of structure

$\sim\sim\sim CH_2CH_2^+$.

The concentration of such ions is low, in consequence of rearrangement reactions.

6. In the event of a leak, air could be drawn into the cracking tubes and an explosive mixture might be formed.

CHAPTER 3

3. $C_2H_4 \xrightarrow[\text{(ii) } HSO_4^-]{\text{(i) } H^+} C_2H_5SO_4H.$ $C_2H_5SO_4H$ is strongly acidic, so

$C_2H_4 + C_2H_5SO_4H \longrightarrow C_2H_5^+ + C_2H_5SO_4^- \longrightarrow EtSO_4Et$

4. $C_2H_4 \xrightarrow{H^+} C_2H_5^+ \xrightarrow{C_2H_4} C_4H_9^+ \xrightarrow{\text{etc.}} C_nH_{2n+1}^+ \xrightarrow{-H^+} C_nH_{2n}$

$C_2H_5^+ \xrightarrow{C_2H_5OH} C_2H_5\overset{+}{\underset{\underset{H}{|}}{O}}C_2H_5 \xrightarrow{-H^+} C_2H_5OC_2H_5$

5. (i) £238 per tonne, (ii) £214 per tonne.

6. No, since ethane would accumulate in the recycle loop.

7. $CH_3CHO \xrightarrow{OH^-} {}^-CH_2CHO + H_2O$

$CH_3CHO + {}^-CH_2CHO \longrightarrow CH_3\overset{\overset{\displaystyle O^-}{|}}{C}HCH_2CHO$

$\xrightarrow{+H^+} CH_3\overset{\overset{\displaystyle OH}{|}}{C}HCH_2CHO$

$CH_3\overset{\overset{\displaystyle OH}{|}}{C}HCH_2CHO \xrightarrow{+H^+} CH_3\overset{\overset{\displaystyle \overset{+}{O}H_2}{|}}{C}HCH_2CHO$

$\xrightarrow{-H_2O} CH_3\overset{+}{C}HCH_2CHO \xrightarrow{-H^+} CH_3CH{=}CHCHO$

8. Oxychlorination.

CHAPTER 4

1. About 240 000 tonnes per annum. Could be adjusted somewhat

by varying the cracking severity. Also, would vary to some extent with the composition of the naphtha used.

3. Di-isopropyl ether, low polymers of propylene.

4. Removal of secondary, benzylic hydrogen to give a resonance stabilised radical is easier than removal of methyl hydrogen. Conditions are fairly selective in contrast to, for example, thermal cracking.

5.

$$\underset{\text{PhCHCH}_3}{\overset{\overset{\displaystyle \text{O}\cdot}{|}}{}} \xrightarrow{\;-[\text{H}\cdot]\;} \underset{\text{PhCCH}_3}{\overset{\overset{\displaystyle \text{O}}{\|}}{}}$$

6. See answer to question 7, Chapter 3.

7.
$$\text{CH}_2{=}\text{CH}_2 \xrightarrow{\text{Oxo process}} \text{CH}_3\text{CH}_2\text{CHO} \xrightarrow{\text{H}_2} \text{CH}_3\text{CH}_2\text{CH}_2\text{OH}$$

CHAPTER 5

1. About 80 000 tonnes per annum.

2. No. Cracker feedstock is ethane.

3. Multistage, consumes chlorine and sodium hydroxide.

4. Inductive effect of two methyl groups, intermediate carbocation more stable.

5.
$$(\text{CH}_3)_2\text{C}{=}\text{CH}_2 \xrightarrow{\text{H}^+} (\text{CH}_3)_3\text{C}^+ \xrightarrow{\text{CH}_3\text{OH}} (\text{CH}_3)_3\overset{+}{\text{C}}\underset{\overset{\displaystyle |}{\text{H}}}{\text{O}}\text{CH}_3$$

$$\xrightarrow{-\text{H}^+} (\text{CH}_3)_3\text{COCH}_3$$

6. (i) Chlorohydrin process or Halcon process, *not* direct oxidation; (ii) ammoxidation.

CHAPTER 6

1. TNT manufacture.

2. Ammonia poisons the acid sites. (i) No aromatics at all; (ii) would be formed from cyclohexane and homologues only.

3. From the aromatic hydrocarbons present in the naphtha, with some loss of side chains, and from secondary reactions.

6. No importance, since all ethylene manufacture based on ethane.

CHAPTER 7

1. Pure cyclohexane cannot be separated from naphtha since there are a number of other components which have similar or identical boiling points.

2. Hydrogen supply.

3.

$$O=CH(CH_2)_4CH_2OO \cdot \longrightarrow O=CH(CH_2)_4CH_2OOH \longrightarrow$$

$$O=CH(CH_2)_4CH_2O \cdot \longrightarrow O=CH(CH_2)_4CHO \xrightarrow[\text{oxdn.}]{\text{further}} HO_2C(CH_2)_4CO_2H$$

$$O=CH(CH_2)_4CH_2 \cdot \longrightarrow O=CH(CH_2)_4CH_3 \xrightarrow[\text{oxdn.}]{\text{further}} CH_3(CH_2)_4CO_2H$$

5.

6.

CHAPTER 8

1. Deactivating effect of nitro group.

3. Operates at high conversion. Also, yield may be a little higher. Corrosive reaction mix increases capital cost.

4. No intermolecular hydrogen bonding in ester.

6. To achieve temperature required.

7. Dehydration, ether formation.

CHAPTER 9

2. $CH_4 + 2H_2O \longrightarrow CO_2 + 4H_2$

   naphtha $\quad + 2H_2O \longrightarrow CO_2 + 3.05H_2$
   (approx. $CH_{2.1}$)

   i.e. more carbon dioxide to remove per mole of hydrogen.

4. About 48%.

5. Transport costs, as aqueous solution, are high. About 1.7 tonnes of water has to be transported with every tonne of formaldehyde.

CHAPTER 10

2.

$$CH_3\overset{\overset{\displaystyle O\cdot}{|}}{C}HCH_2CH_3 \xrightarrow{-[H\cdot]} CH_3\overset{\overset{\displaystyle O}{\|}}{C}CH_2CH_3$$

$$CH_3CH_2CH_2CH_3 \xrightarrow{-[H\cdot]} CH_3CH_2CH_2CH_2\cdot$$

$$\longrightarrow CH_3CH_2CH_2CH_2OO\cdot \longrightarrow CH_3CH_2CH_2CH_2OOH$$

$$\longrightarrow CH_3CH_2CH_2CH_2O\cdot \longrightarrow HCHO + CH_3CH_2CH_2\cdot$$

$$\longrightarrow CH_3CH_2CH_2OOH \longrightarrow CH_3CH_2CH_2O\cdot$$

$$CH_3CH_2CHO \xrightarrow{\text{further oxidation}} CH_3CH_2CO_2H$$

4. Not feasible to make tube that would withstand the temperatures and transmit heat at the rate required.

Index

acetaldehyde, 55–59
  from acetylene, 166
  oxidation, 58
  use, 137
acetic acid, 151–152, 163–164
  as solvent for oxidation, 129, 136, 137
  by carbonylation of methanol, 151–152
  by oxidation of acetaldehyde, 58–59
  by oxidation of butane or naphtha, 163–164
  uses, 52, 111, 152, 154
  vinyl acetate from, 59–60
acetic anhydride, 59, 152, 154
acetone, 67–68
  as co-product, 116, 117, 164
  bisphenol A from, 119
acetonitrile, 78, 81
acetophenone, 116
acetylene, 166–167
  acetaldehyde from, 55
  acrylic acid from, 75
  acrylonitrile from, 76
  chloroprene from, 83
  vinyl acetate from, 59
  vinyl chloride from, 47, 48
acrylonitrile, 76–78
  in nitrile rubber, 83
adiabatic reactors, 24, 25
adipic acid, 109
  in cyclohexane oxidation, 107
  manufacture of, 82, 109
  nylon 66 from, 110–111
Acrilan, 78

acrolein, 75, 76, 77, 78
acrylic acid, 75–76
acrylic fibres, 78
adiponitrile, 109–110
  from acrylonitrile, 78
  from butadiene, 84
aldol, condensation, 59
l-alkenes, linear, 61–62
alkyd resins, 134
alkylation, 102–103, 116, 123–124
alkylbenzenesulphonate detergents, 123–124
allyl alcohol, 79
allyl chloride, 78-79
ε-aminocaproic acid, 112
Amoco process, 136–137
ammonia
  manufacture, 143–147
  oxidation, 148
  uses, 52, 77, 109, 111, 147–148, 152–153
ammonium nitrate, 147
ammonium sulphate, 112, 148
ammoxidation, 77–78
aniline, 120–122
antifreeze, 52
aromatic hydrocarbons (*see also* benzene,
    toulene, xylenes)
  cracking of, 39
  manufacture, 21, 89–100, 154
  market demand, 96
  separation, 93–95, 97
associated gas, 16
atactic polymers, 65

atmospheric residue, 18, 19, 158

benzene
 alkylation, 21–22, 101–103, 116, 124
 chlorination, 114
 derivatives, 101–124
 hydrogenation, 105–106
 manufacture, 89–96, 99–100, 154
 nitration, 120–121
 oxidation, 113, 122
 by-product in ethylbenzene dehydrogenation, 104
 sulphonation, 114
benzoic acid, 129–130
benzenesulphonic acid, 114
bis(2-hydroxyethyl) terephthalate, 138
bisphenol A, 119
butadiene, 81–85
 formation in cracking, 34, 40
butane
 as ethylene feedstock, 30
 dehydrogenation, 82
 in natural gas, 16, 17
 oxidation, 122, 163–164
butenes, formation in cracking, 19, 34, 40
n-butenes
 dehydrogenation, 82
 hydration, 87
 oxidation, 122
l-butene, 47
butanal see butyraldehyde
butanol see butyl alcohol
n-butyl alcohol, 59, 73, 74
s-butyl alcohol, 87
t-butyl alcohol, 70, 71
t-butyl hydroperoxide, 69, 70
n-butyraldehyde, 72, 73, 74
butylene see butene
butyl rubber, 85

capital cost, 26, 27, 28
caproic acid, 107
caprolactum, 111–113, 130
carbon dioxide, scrubbing, 53, 145
carbon disulphide, 162–163
carbon monoxide
 in acetic acid process, 151
 in acrylic acid process, 75
 in Oxo process, 72–74, 155
 in stem-reforming, 141–146, 148–149,
  155–156
 manufacture, 156
 uses, 82, 128, 154
carbon tetrachloride, 159–162, 163
carbonyl chloride, 122, 128
p-carboxybenzaldehyde, 136
catalytic cracking, 19, 20
catalytic cracker gases, 19, 20
catalytic reactors, 24, 25
catalytic reforming, 19, 21, 90–95
catalytic reformate, 19
cellulose acetate, 152

chlorination
 benzene, 114
 butadiene, 83–84
 1,1-dichloroethane, 60
 ethane, 61
 ethylene, 47, 48
 methane, 159–161
 propylene, 78–79
 linear alkanes, 165
chlorinolysis, 161
chlorobenzene, 114, 128
2-chloro-1,3-butadiene, see chloroprene
chlorodifluoromethane, 162
chloroethane, see ethyl chloride
chloroethene, see vinyl chloride
chloroform, 159–162
chloromethane, see methyl chloride
chloroprene, 83–84
coal carbonisation, 89, 90, 132
coal gasification, 156–157
coal tar, 113
continuous processing, 21–25
conversion, definition, 21–22
co-ordination polymerisation, 45–46
costs, 26–28
Courtelle, 78
cracking, acetylene by, 166–167
cracking, ethylene and co-products by, 21,
  30–42
 ethane, 33, 34–35
 first use, 14
 naphtha and gas oil, 33, 34, 36–40
 product distribution, 34
 propane, 33, 34, 35–36
 secondary reactions, 39–41
 severity, 40
crude oil (petroleum), 13, 15–16, 17–20
crystalline regions, 46
cumene, 72, 115–117
cumene hydroperoxide, 115, 116, 117
cumene process for phenol, 72, 115–117
cyanobutenes, 85
cycloalkanes, cracking of, 39
cyclohexane, 105–108, 115
cyclohexanol, 106–108
 dehydrogenation, 111
 oxidation, 109
 phenol from, 115
cyclohexanone, 106–108
 caprolactam from, 111–112
 from cyclohexanol, 111
 from phenol, 119–120
 oxidation, 109
 phenol from, 115
cyclohexyl hydroperoxide, 106, 107, 108

dehydrocyclisation, 91–92
dehydrogenation
 n-butane, 82
 n-butenes, 82
 s-butyl alcohol, 87

cyclohexanol, 111
ethanol, 55
ethylbenzene, 101, 104
in catalytic reforming, 91
isopropyl alcohol, 67
linear alkanes, 165
methanol, 150–151
dehydroisomerisation, 91, 92
detergent alkylate, 123–124
detergents, 123–124
dialkyl phthalates, 133
4,4'-diaminodiphenylmethane, 121—122
dibutyl phthalate, 74
o-dichlorobenzene, 128
1,4-dichloro-2-butene, 83, 84
3,4-dichloro-1-butene, 83, 84
dichlorodifluoromethane, 162
1,1-dichloroethane, 60
1,2-dicloroethane, 14, 47–48
dichloromethane, see methylene chloride
1,4-dicyano-2-butene, 84
diesel fuel, 18
diethylene glycol, 51
dimethyl ether, 153
dimethylformamide, 81
dimethyl terephthalate, 134, 135, 137–138, 153
dinitrotoluene, 127
4,4'-diphenylmethane di-isocyanate (MDI), 122
distillation, 24–26
dodecene, 123
dry natural gas, 16
dual function catalyst, 95

economies of scale, 28
epichlorohydrin, 79
epoxyethane, see ethylene oxide
epoxypropane, see propylene oxide
ethanal, see acetaldehyde
ethane
    chlorination, 61
    cracking, 14, 21, 30–31, 33, 34–35
    in natural gas, 16, 17
1,2-ethanediol, see ethylene glycol
ethanoic acid, see acetic acid
ethanol, 53–55
    acetaldehyde from, 55
    from methanol, 154
    uses, 76, 128
ethanolamines, 52–53, 145
ethene, see ethylene
ethenoic aciod, see acrylic acid
ethyl acrylate, 76
ethyl alcohol, see ethanol
ethyl chloride, 60–61
ethylbenzene, 22, 101–104
    in catalytic reformate and pyrolysis gasoline, 96, 97
    isomerisation, 98
    oxidation, 70

ethylbenzene hydroperoxide, 70
ethylene
    derivatives, 43–62
    feedstocks for, 30–31
    from methanol, 154
    hydration of, 53–54
    hydrochlorination of, 60–61
    in styrene manufacture, 53, 101–103
    manufacture, 14, 21, 30–42
    oxidation, 50, 55–58, 59
    oxychlorination, 48–49
    polymerisation, 44–46
    reaction with sulphuric acid, 54
ethylene chlorohydrin, 14, 49–50
ethylene dichloride, see 1,2-dichloroethane
ethylene glycol, 14, 51–52, 138, 154
ethylene oxide, 49–53, 76
2-ethylhexanol, 74
ethyne, see acetylene

fatty alcohols, 62
fluidised bed reactors, 24, 25
formaldehyde, 150–151
    uses, 118–119, 122, 154
furfural, 81

gas oil
    as petrochemical feedstock, 20, 21, 30–31
    cracking, 33, 34, 36–41
    in refining, 18
gasoline,
    from methanol, 155
    in refining, 18–19
    lead additives for, 61, 161
    methanol in, 154
    methyl t-butyl ether in, 86
    glycerol, 71, 134
    growth reaction, 61, 62

Halcon process for propylene oxide, 69
heavy fuel oil, 18
hexamethylene diamine, 110
hexanedioic acid, see adipic acid
1-hexene, 47
hydration
    acetylene, 55
    n-butenes, 87
    ethylene, 53–55
    ethylene oxide, 51
    propylene, 66–67
    propylene oxide, 71
hydrocracking, 92
hydrodealkylation, 99–100
hydrodesulphurisation, 90, 144
hydroformylation, 72–74
hydrogen, manufacture, 155
hydrogen cyanide, 67, 76, 78, 85
hydrogen sulphide, 17, 53
hydrogenation
    adiponitrile, 110
    aldehydes, 73, 74

benzene, 105–106
benzoic acid, 130
dinitrotoluene, 127, 128
in terephthalic acid purification, 136
nitrobenzene, 121
phenol. 119–120
pyrolysis gasoline, 95
hydroxylamine, 111, 112

isobutane, oxidation, 70
isobutene, 85–86
from *t*-butyl alcohol, 71
isobutyl alcohol, 73, 74
isobutyraldehyde, 72, 73, 74
isoprene, 85
isopropyl alcohol, 14, 66–67
isotactic polymers, 65

kerosine, 18, 20, 165
ketene, 152

light gasoline, 18, 19
linear alkanes, 165
linear alkenes, 123–124, 165, 166
linear alkylbenzenesulphonates (LABS), 123–124

MDI (4,4′-diphenylmethane di-isocyanate), 122
MTBE (methyl *t*-butyl ether), 86, 153
maleic anhydride, 87, 122–123, 133
methanal, *see* formaldehyde
methanation, 145
methane (*see also* natural gas)
acetylene from, 167
carbon disulphide from, 162–163
chlorination, 159–161
in natural gas, 16, 17
steam reforming, 21, 141–146, 148–149, 155–156
methanol, 148–155
methyl *t*-butyl ether from, 86
methyl chloride from, 161
uses, 67, 82, 110, 128, 135, 138
methyl *t*-butyl ether, 86, 153
methyl chloride, 153, 159–162
methyl ethyl ketone, 87
methyl isobutyl ketone, 68
methyl methacrylate, 67, 68, 153
methylamines, 152—153
methylene chloride, 159–162
2-methylpropanal, *see* isobutyraldehyde
2-methyl-1-propanol, *see* isobutyl alcohol
2-methylpropene, *see* isobutene
N-methylpyrrolidone, 81
methylsilanes, 161
methylstyrene, 116
monochlorobenzene process for phenol, 114

naphtha
as petrochemical feedstock, 20, 21, 30–31

catalytic reforming, 21, 90–95
cracking, 21, 33, 34, 36–41
in refining, 18
oxidation, 163–164
steam reforming, 141, 142, 143, 144
naphthalene, 132
naphthenes, 16
natural gas, 15, 16–17, 30–31
neoprene, 84
nitration
benzene, 120–121
toluene, 127, 131
nitric acid
manufacture, 148
uses, 109, 120, 121, 127, 131, 135
nitrile rubber, 83
nitrobenzene, 120–121
nitrotoluene, 127
nonylphenol, 53
nylon, 6, 112–113
nylon, 66, 110–111
nylon salt, 110

octane number, 18–19
oil refining, 17–19
operating cost, 27
Orlon, 78
outline flow scheme, 22
oxidation
acetaldehyde, 58–59
ammonia, 148
benzene, 113, 122
benzoic acid, 130
n-butane, 122, 163–164
n-butenes, 87, 122
cyclohexane, 106–108
cyclohexanol/cyclohexanone, 109
cumene, 116
ethylbenzene, 70
ethylene, 50, 55–58, 59–60
isobutane, 69–70
methanol, 150–151
naphtha, 163–164
naphthalene, 131–132
propylene, 67, 75–76
toluene, 129
o-xylene, 131–132
p-xylene, 134–137
Oxo process, 72–74, 75, 133, 155
oxychlorination, 48–49, 60, 114

paraffin wax, 165–166
paraffins, 15
partial oxidation, 157–158
perchloroethylene, 60, 161
Perspex, 68
petrochemicals
definition, 13
early development, 14
petrol, *see* gasoline
petroleum, 13, 15–16, 17–20

petroleum refining, 17–20
phenol-formaldehyde resins, 117–119
phenol, 113–120, 130
1-phenylethanol, 70, 71
2-phenyl-2-propanol, 116
phthalic anhydride, 131–134
plasticisers, 49, 133
Plexiglas, 68
primary alcohols, linear, 62
production cost, 26, 27, 28
propane
　as feedstock, 30–31
　cracking, 21, 33, 34, 35–36
　in natural gas, 16, 17
2-propanol, *see* isopropyl alcohol
propene, *see* propylene
propylene
　ammoxidation, 77–78
　chlorination, 78–79
　chlorinolysis, 161
　derivatives, 64–79
　first petrochemical application, 14
　formation in thermal cracking, 34, 36, 40, 42
　hydration, 66–67
　in cumene process, 116
　manufacture and sources, 19, 21, 34, 41–42, 64
　oxidation, 67, 75–76
　polymerisation, 65
propylene chlorohydrin, 68
propylene glycol, 71–72, 133
propylene oxide, 28, 68–72
propylene tetramer, 74–75, 123
propylene trimer, 74–75
polybutadiene rubber, 83
polychloroethane, *see* poly(vinyl chloride)
polychloroprene (neoprene), 84
polyester resins, 123, 133–134
polyethylene, 44–47
polyethylene glycols, 51
poly(ethylene terephthalate), 134, 138–139
poly(methyl methacrylate), 68
polyols, 53, 71
polypropylene, 65
polystyrene, 105
poly(vinyl chloride), 49, 133
polyurethanes, 126, 128–129
pressure swing adsorption, 155
pyrolysis gasoline, 34, 42, 95–96

Raschig process for phenol, 114–115
reactors, 23–24, 25
recycling, 22
refining, 17–20

SBR (styrene-butadiene rubber), 82–83, 105
Scientific Design process for phenol, 115
severity, 40
shift reaction, 142, 144

single cell protein, 155
solid phosphoric acid, 54, 66, 74, 103, 116
steam cracking, *see* cracking
steam reforming, 21, 141–146, 148–149, 155–156
stirred flow reactors, 23
styrene, 101–105
　co-product of propylene oxide, 71
　ethylene use in, 53
　outline flow scheme, 21
　uses, 83, 134
styrene-butadiene rubber, 82–83, 105
sulfolane, 95
sulphonation, 114, 124
sulphonation process for phenol, 114
synthesis gas, definition, 141
synthetic rubbers, 82–83, 84

TDI (tolylene di-isocyanate), 126–129
TNT (trinitrotoluene), 131
terephthalic acid, 134–139
tetrachloromethane, *see* carbon tetrachloride
tetraethylene glycol, 95
tetraethyl-lead, 61
tetramethyl-lead, 161
toluene
　by-product in styrene manufacture, 22, 104
　derivatives, 126–131
　disproportionation, 99, 100
　hydrodealkylation, 99–100
　manufacture, 89–96
　nitration, 127, 131
　oxidation, 129
p-toluic acid, 134, 135, 136
tolylene diamine, 127
tolylene di-isocyanate, 126–129
transalkylation, 102–103
tributylphosphine, 73
1,1,1-trichloroethane, 60
trichloroethylene, 60, 166
trichlorofluoromethane, 162
trichloromethane, *see* chloroform
triethylaluminium, 61, 62
trinitrotoluene, 131
triphenylphosphine, 73
tubular reactors, 23, 24, 25

unsaturated polyesters, 105, 123, 133–134
urea, 148

vinyl acetate, 59–60, 152, 154
vinyl chloride, 47-49, 60
vinylacetylene, 83, 84

Wacker Chemie process, 55–58
water gas reaction, 156
wax cracking, 165–166
wet natural gas, 16, 21, 30, 141
Witten process, 135

o-xylene, phthalic anhydride from, 131–133

p-xylene, terephthalic acid and dimethyl terephthalate from, 134–137

xylenes, 89–99, 100
   in catalytic reformate and pyrolysis gasoline, 96
   isomerisation, 98–99
   separation, 97

yield, definition, 21–22

Ziegler-Natta polymerisation, 45–46, 65, 83